LIVING MATTER

Algebra of Molecules

LIVING MATTER

Algebra of Molecules

Valery V. Stcherbic
Taras Shevchenko National University
Kiev, Ukraine

Leonid P. Buchatsky
Taras Shevchenko National University
Kiev, Ukraine

CRC Press
Taylor & Francis Group
Boca Raton London New York

CRC Press is an imprint of the
Taylor & Francis Group, an **informa** business
A SCIENCE PUBLISHERS BOOK

CRC Press
Taylor & Francis Group
6000 Broken Sound Parkway NW, Suite 300
Boca Raton, FL 33487-2742

First issued in paperback 2020

ISBN-13: 978-1-4987-4137-8 (hbk)
ISBN-13: 978-0-367-73753-5 (pbk)

Library of Congress Cataloging-in-Publication Data

Shcherbik, V. V.
Living matter : algebra of molecules / Valery V. Stcherbic, Leonid P. Buchatsky.
pages cm
"A CRC title."
Includes bibliographical references and index.
ISBN 978-1-4987-4137-8 (hardcover : alk. paper) 1. Biochemistry. 2. Biophysics. 3. Matter--Constitution. 4. Clifford algebras. I. Buchatskii, L. P. II. Title.

QP514.2.S625 2015
572--dc23 2015028930

Visit the Taylor & Francis Web site at
http://www.taylorandfrancis.com

and the CRC Press Web site at
http://www.crcpress.com

Preface

Living matter is built out of "nonliving molecules" and obeys the laws of physics and mathematics. In living organisms, activation of chemical substances takes place: biomolecules possess complex structure and fulfil specified functions.

Living matter is a highly organized matter. Biochemical reactions, which maintain self-organization of living matter, occur within a limited range of environmental parameters. These reactions occur due to enzymes. Highly coordinated biological processes determine the development of organisms from a single cell. Biochemical processes obey the laws, appropriate to both the most simple and the most highly developed organisms. Unity of living matter is in the property of self-reproduction. Its basis is DNA, the main molecule of life. DNA codes and genes, completely determine the structure of the living organism.

A mathematical basis for building biological structures and biochemical processes may someday be found. The laws of physics and chemistry are quite universal but among them there is no the law of biological activation of nonliving molecules. The emergence of life from nonliving matter and the evolution of living organisms are some of the most complicated questions standing before humankind.

Theoretical physics is an exact science with a high number of mathematical approaches and laws that describe the surrounding world. In biology, there are many "exceptions to the rules"; even genetic code has variations. High accuracy of biological processes is the basis for the emergence of various types of symmetries observed in structural molecular biology. Forms of symmetry are often concealed in a mathematical formula and cannot be seen to the observer of biological structures. For each symmetry, there is an adequate law that represents maintenance of some physical value in time. Application of algebraic methods can reveal essential features of biological processes that are difficult to derive from differential equations.

This reasoning is the basis for using Clifford algebra to describe structures and processes in living matter. We do not know why Clifford algebra may be the reason for biological activation of nonliving molecules, but numerous examples considered in this book, show that Clifford algebra possesses universality for adequate description of both biological structures and biochemical processes.

The simplicity of the emergence of Clifford algebra is amazing. Any section in the sequence of the growing cell number is a set of Clifford algebras. If a geometrical structure possesses mirror symmetry (e.g., butterfly's wings), it is a spinor group of some Clifford algebra. But asymmetrical geometrical structure also is connected with some Clifford algebra, which is not so obvious.

Concepts of Clifford algebra play the leading part in this book; less attention has been paid to internal symmetries of the algebra itself, e.g., to spinor groups. Main attention is paid to those structural elements of biomolecules that determine a specific Clifford algebra.

The authors hope that the book will be interesting to specialists in theoretical biology, quantum genetics, biophysics and bioinformatics and would appreciate any remarks and suggestions.

March 2015

Valery V. Stcherbic
Leonid P. Buchatsky

Contents

1

DNA Structure and Clifford Algebra

Deoxyribonucleic acid (DNA) takes central place in storage of genetic information. DNA molecule is a polymer consisting of two oppositely directed chains, which are twisted in the form of a right-handed double helix. Each chain is built from nucleotide units that contain sugar-phosphate group and nitrogenous bases, nucleotides that are encoding genetic elements. The DNA structure contains four nucleotides: adenine **A**, guanine **G**, cytosine **C** and thymine **T**. The Sugar-phosphate group consists of 2-deoxyribose and phosphoric acid residues. DNA chain orientation is identified by carbon atoms of 2-deoxyribose: **(5′)CH**$_2$ and **(3′)COH**.

The biological function of DNA and storage and transfer of genetic information to daughter cells is based on specific, complimentary pairing of nucleotides: **A** is paired with **T**, and **G**—with **C**. Pairs **AT** and **GC** are nonpolar and almost equal in mass and length. Complimentary pairing of nucleotides is the basis of semiconservative DNA replication.

Ribonucleic acid (RNA) can also store genetic information. A single RNA helix is seldom used as a carrier of genetic information (only in some viruses); its main role is storing DNA sites as copies of individual protein-coding genes (mRNA) or in formation of large structural complexes, e.g., ribosomes and spliceosomes. At self-splicing, RNA may perform the function of an enzyme. RNA also performs an important role during DNA replication. So called RNA-primers are necessary to synthesize DNA complementary chains, although this fact is not obvious. RNA contains sugar, ribose, which hydroxyl groups make more reactive than DNA. Besides, RNA contains uracil **U**, which is somewhat lighter than thymine.

DNA REPLICATION

DNA replication is a complicated biochemical process. DNA is a very long molecule and it is rather strange that such a small number of enzymes take part in the reproduction of genetic information. DNA is the chief molecule of living matter and its conservation over millions of years seems unbelievable.

A lot of effort has been made by physicists and mathematicians (not mentioning biologists) to elucidate chief moments of DNA replication. Why, in general, is the process of DNA duplication possible? On experimental material, biologists have shown that DNA replication is impossible without complementarity; however theoretician-biologists cannot clearly understand why such a complex mechanism of replication is necessary and how it could appear during evolution. Atoms of a crystalline structure find their place quite simply. And this fact haunts biophysicists: why cannot quantum mechanics explain the process of DNA replication?

We still know little about the structure and properties of the space wherein molecules of living cells "live". Biologically active molecules are almost always complex and polyatomic; nevertheless, they selectively interact with other molecules, that is, they "recognize" them. Nucleotides possess the intrinsic reserve of energy, which is sufficient to proceed with elongation of the synthesizing DNA chain. Rate of DNA synthesis is high in bacteria (1000 np/s), but less in eukaryotes (100 np/s). Note that this is a rather selective process, not simple polymerization. Matrix synthesis, a characteristic feature of living matter, is the basis for DNA replication.

Another aspect: information is the basis of life. Carriers of genetic information in a living cell are nucleic acids and amino acids. They are also transformers of the information and the most important of them are ribosomes, nucleosomes, spliceosomes, as well as numerous enzymes. Information is an abstract notion, which has a formal numeric representation. DNA alphabet contains only four letters—**A**, **T**, **G**, **C**. Could this be connected with the four-dimensionality of our world? Generally, biophysicists cannot see any reasons to use theory of relativity for characterization biological processes. Therefore, at best, advances have been made to operate quantum-mechanical notions instead of classical reasoning on electrostatic forces. Arguments of spins of separate molecules lead to abstract binary codes. There may be made infinite number of codes for complementary pairs **AT** and **GC**; significantly more difficult is to suggest the content of codes for separate nucleotides.

One more issue concerning DNA replication is condensation of chromosomes. At any number presentation suggested for DNA, there always may appear the idea of pressing informational content of DNA.

Points belonging to the space of DNA states should condense; this process becomes apparent not only mechanically in DNA compacting, but, as we suppose, is connected with compression of DNA information. One may imagine the picture of compressed spring preceding the next generation of cells during the growth and formation of an organism.

The problem of mechanicality: In spite of using different mathematical apparatus, theoretical physicists could not get rid of the mechanical approach to natural phenomena. Lagrange's function, gauge theory of field, particles and antiparticles—all this is too far from biological processes.

CLIFFORD ALGEBRA

All attempts to understand the phenomenon of DNA replication (without enzymes) while not applying the theory of relativity were unsuccessful. Four-dimensional space-time is pseudo-Euclidean. This means that the square length of Lorentz interval contains both positive and negative squares that may be of two types:

$$\text{I: } s^2 = t^2 - x^2 - y^2 - z^2 \text{ or II: } s^2 = -t^2 + x^2 + y^2 + z^2.$$

The first type of **Lorentz intervals I** is denoted as [1+ 3–], the second type of **Lorentz intervals II**—as [3+ 1–].

Semiconservative mode of DNA replication suggests that squares of Lorentz intervals may be divided by two and doubled. Using Clifford algebra, from two squares, four squares may be obtained! This idea is key towards the explanation of DNA replication.

Further, when explaining DNA structure, we will take advantage of numerous properties of Clifford algebra. In contrast with standard Clifford algebra, where the unit element e_0 is passive (not generator), we will often use e_0 as an active element of quadratic form; that is, the unity e_0 differs from the simple unity. Lorentz intervals cannot be divided without this modification of Clifford algebra.

Since vectors of Clifford algebra do not commutate between themselves (except e_0), it is clear that we have to deal with **noncommutative geometry** of space-time. Why, in the frame of such geometry, is it possible to realize synchronous movements of a large number of particles? To confirm this hypothesis, further investigations are needed.

Moreover, application of Clifford algebra to explain the DNA structure is also connected with the fact that opposite properties of biological structures may be juxtaposed with positive and negative squares of Clifford algebra elements; even chemical elements may be considered as generators (**charges**) of Clifford algebra. This may resolve the problem of antiparticle absence in biological structures.

DNA REPLICATION AND CLIFFORD ALGEBRA

Clifford algebra $Cl(2)$ with two generators contains four vectors, squares of which determine Lorentz intervals I or II with noncommutative coordinates.

Let us consider reproduction of intervals [3+ 1–] without involving experimental data on DNA replication.

Thus, we should double the interval square

$$s^2 = -t^2 + x^2 + y^2 + z^2.$$

There is only one possibility: any two squares should be considered as generators of Clifford algebra—these are algebras $Cl(2, 0)$ and $Cl(1, 1)$. Both algebras are isomorphic and generate new intervals [3+ 1–]:

$$s^2 = (-t^2 + x^2) + (y^2 + z^2) \rightarrow Cl(1, 1) \oplus Cl(2, 0) \rightarrow$$
$$\rightarrow (-t_1^{\,2} + x_1^{\,2} + y_1^{\,2} + z_1^{\,2}) \oplus (-t_2^{\,2} + x_2^{\,2} + y_2^{\,2} + z_2^{\,2}).$$

Here, time coordinates refer to the bases e_{12}, $(e_2)_1$, $(e_{12})_2$. In algebra $Cl(1, 1)$, time basis vector is a generating one, whereas in algebra $Cl(2, 0)$, it is generated by coordinate basis vectors. Recall that in the Dirac electron theory, time basis vector is a generator of Clifford algebra $Cl(1, 3)$.

So, why is the initial interval [3+ 1–] doubled? We postulate that **the property of space-time intervals to multiply is natural for noncommutative geometry**.

The algebras are isomorphic, consequently, the intervals are also isomorphic but may refer to different biochemical processes, that is, the intervals are not strict copies.

Clifford algebra consists of the assembly of Lorentz intervals of I and II types. If we reject some special intervals associated with the centre of Clifford algebra, then we could suppose that internal Lorentz intervals of the type I are identical between themselves, and Lorentz intervals of the type II are also identical between themselves. Identity of Lorentz intervals belonging to different Clifford algebras is, possibly, non-essential for mathematics, but it is important for biology, which considers an enormous number of interacting molecules. Lorentz intervals with noncommutative variables may be reduced to ordinary Lorentz intervals.

Initial Lorentz interval [3+ 1–] transformed into its two isomorphic copies: the old interval disappears but it produced two new intervals. This is just a naive approach to the multiplication of Lorentz intervals, but DNA does not divide in this way!

Each DNA chain is a matrix for synthesis of a complementary chain—initial chains are not destroyed. We will assume that initial Lorentz interval [3+ 1–] is **nonlocally** a matrix for synthesis of its copies. In this case this interval is not destroyed. We have only to pass from nonlocal matrix to the local one. If this transition exists, the local complementarity of DNA chain will be equivalent to its nonlocal copies.

Let us suppose that DNA chains consist of Lorentz intervals of Clifford algebras; at that, one chain **locally** contains only intervals [3+ 1–], while the complementary chain contains only intervals [1+ 3–].

It has been found, experimentally, that leading **chain 5′3′** (the beginning of replication) is synthesised complementary to the **chain 3′5′**. Leading chain always follows helicase. The **chain 5′3′** is copied **nonlocally** earlier, while the **chain 3′5′**—later, as it needs transformation of intervals. Copying of DNA chains goes by inserting RNA primers with Lorentz intervals [3+ 1–]. DNA nucleotides are attached to 3′-end of 2-deoxyribose.

Figure 1.1 shows diagrams of division of Lorentz intervals; they are considered as nonlocal matrices in terms of information obtained from experimental data.

Division of Lorentz interval [1+ 3–] does not lead to its duplication. RNA primers in both leading and lagging chains consist of Lorentz intervals [3+ 1–]. Okazaki fragments in the lagging chain are a mixture of Lorentz intervals of the type I and II (Figure 1.2).

Terminal RNA primers on the side of **5′**-end of synthesized DNA chain are moved off by DNA polymerase, while internal RNA primers of Okazaki fragments are substituted for DNA parts.

Inversion of Lorentz interval [3+ 1–] is necessary only for RNA primers in Okazaki fragments. To do this in Clifford algebra $Cl(1, 1)$, it takes only to use the inversion of one vector $e_1 \rightarrow ie_1$. Just the inversion operation of type II Lorentz intervals causes the lagging chain to be synthesized discretely.

Initial DNA consists of two chains. As soon as DNA chains are untwisted by helicase, synthesis of leading chain begins. Transition from nonlocal

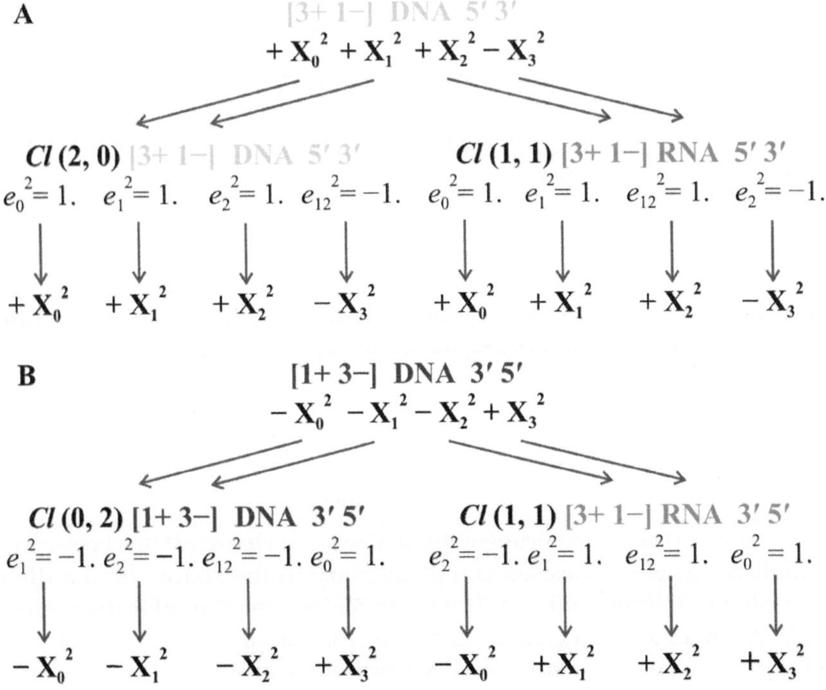

Figure 1.1 Division of DNA Lorentz intervals.
(**A**) Leading chain. (**B**) Lagging chain.

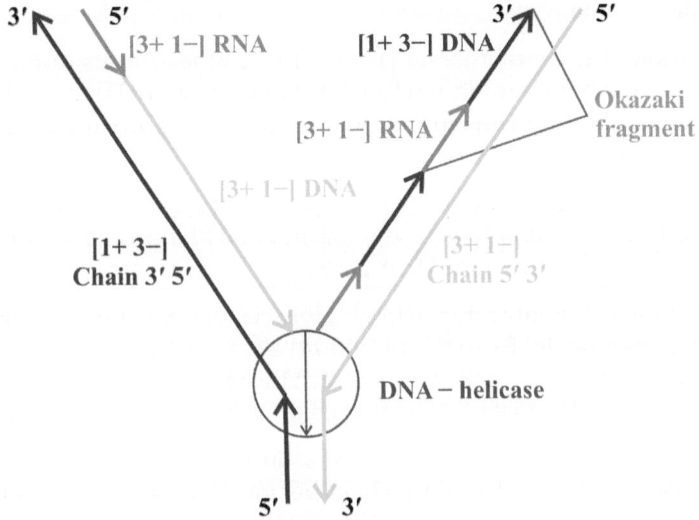

Figure 1.2 Diagram of DNA replication.

DNA matrix (**chain 5′3′**) to the local one (**chain 3′5′**) is carried out with the use of Clifford algebra

$$Cl(2, 1) = [3+\ 1-]\ \text{RNA}\ (+)\ [3+\ 1-]\ \text{DNA}$$

Why exactly the algebra $Cl(2, 1)$? The answer is: two old untwisted DNA chains and one new synthesizing leading chain determine **charges (signature of generators)** exactly of the algebra $Cl(2, 1)$.

Synthesis of the lagging chain is determined by the algebra

$$Cl(3, 0) = [3+\ 1-]\ \text{RNA}\ (+)\ [1+\ 3-]\ \text{DNA}$$

where three DNA chains (two old and one leading new chain) determine **charges** of just the algebra $Cl(3, 0)$.

Initial DNA chain, non-local for the leading chain, is the old **chain 5′3′** and return to the synthesis of the lagging chain is a local operation; therefore, negative part of the Clifford algebra signature is equal to zero.

On the contrary, firstly the algebra $Cl(1, 2)$, and then the algebra $Cl(0, 3)$ cannot occur, since the algebra $Cl(0, 3) = [1+\ 3-] \oplus [1+\ 3-]$ is not algebra of the Lorentz interval reproduction.

Another variant—at first the algebra $Cl(1, 2)$, and then the algebra $Cl(3, 0)$—is also impossible, because $Cl(1, 2)$ is isomorphic to $Cl(3, 0)$, but synthesis of leading chain differs from synthesis of lagging chain.

Further, we present a description of key biochemical processes, which occur at DNA replication, on the basis of formulae for transforming elements of Clifford algebras. We use the transformation of generator signature into the signature of all elements and vice versa and rearrangement of elements of Clifford algebras.

- Elongation of leading

$$\underset{1}{[1+\ 3-]\ \text{DNA}}\ (+)\ \underset{2}{[3+\ 1-]\ \text{DNA}} \rightarrow \underset{3}{[2+\ 2-]} \rightarrow \underset{4}{Cl\ (2, 2)} \rightarrow$$
$$\rightarrow \underset{5}{Cl(3, 1)} \rightarrow \underset{6}{[3+\ 1-]\ \text{DNA}}$$

and lagging

$$\leftarrow \underset{4}{[0+\ 3-]} + \underset{3}{[3+\ 0-]} \leftarrow \underset{1}{[1+\ 3-]\ \text{DNA}}\ (+)\ \underset{2}{[3+\ 1-]\ \text{DNA}}$$
$$\leftarrow \underset{8}{[2+\ 6-]\ (+)} \underset{7}{[4+\ 4-]} \leftarrow \underset{6}{Cl(0, 3)}\ (+)\ \underset{5}{Cl\ (3, 0)} \leftarrow$$
$$\underset{10}{[1+\ 3-]\ \text{DNA}} \leftarrow \underset{9}{Cl(1, 3)} \leftarrow$$

DNA chains is joined by Clifford algebra $Cl(1, 3) = [6+\ 10-]$.

- RNA primer and DNA in leading chain, consisting of the type II Lorentz intervals, induce synthesis of RNA primer and DNA in lagging chain, that is, the mixture of Lorentz intervals of types I and II:

$$[3+\ 1-]\ \text{RNA}\ (+)\ [3+\ 1-]\ \text{DNA} = Cl\ (2, 1) \to [2+\ 2-] + [4+\ 0-] \to$$
$$\to Cl\ (2, 2)\ (+)\ Cl\ (4, 0) \to [10+\ 6-] + [6+\ 10-] \to$$
$$\to Cl\ (3, 1)\ (+)\ Cl\ (1, 3) \to [3+\ 1-]\ \text{RNA}\ (+)\ [1+\ 3-]\ \text{DNA} = Cl\ (1, 2)$$

- Substitution of [3+ 1–] of RNA primer for [1+ 3–] of DNA in lagging chain (rearrangement of Clifford algebras):

$$[3+\ 1-]\ \text{RNA}\ (+)\ [1+\ 3-]\ \text{DNA} = Cl\ (1, 2) \to [4+\ 0-] + [0+\ 4-] \to$$
$$\to Cl\ (4, 0)\ (+)\ Cl\ (0, 4) \to [6+\ 10-] + [6+\ 10-] \to$$
$$\to Cl\ (1, 3)\ (+)\ Cl\ (1, 3) \to [1+\ 3-]\ \text{DNA}\ (+)\ [1+\ 3-]\ \text{DNA} = Cl\ (0, 3)$$

The transition $Cl(1, 2) \to Cl(0, 3)$ is carried out by the substitution $e_1 \to ie_1$.

- There is reverse conversion of Okazaki fragment into leading chain:

$$[3+\ 1-]\ \text{RNA}\ (+)\ [1+\ 3-]\ \text{DNA} = Cl\ (1, 2) \to [2+\ 2-] + [2+\ 2-] \to$$
$$\to Cl\ (2, 2)\ (+)\ Cl\ (2, 2) \to [10+\ 6-] + [10+\ 6-] \to$$
$$\to Cl\ (3, 1)\ (+)\ Cl\ (3, 1) \to [3+\ 1-]\ \text{RNA}\ (+)\ [3+\ 1-]\ \text{DNA} = Cl\ (2, 1)$$

- We may suppose that remote terminal RNA primers are formally equivalent to RNA with intervals [1+ 3–]. Then substitution by telomerase of remote terminal RNAs for DNAs (transformation of Clifford algebras) is:

$$[1+\ 3-]\ \text{RNA}\ (+)\ [1+\ 3-]\ \text{RNA} = Cl\ (0, 3) \to [2+\ 2-] + [0+\ 4-] \to$$
$$\to Cl\ (2, 2)\ (+)\ Cl\ (0, 4) \to [10+\ 6-] + [6+\ 10-] \to$$
$$\to Cl\ (3, 1)\ (+)\ Cl\ (1, 3) \to [3+\ 1-]\ \text{DNA}\ (+)\ [1+\ 3-]\ \text{DNA} = Cl\ (1, 2)$$

We considered main moments of DNA replication supposing that one of the DNA chains strictly consists of one type of Lorentz intervals; namely, leading chain always consists of Lorentz intervals [3+ 1–] irrespective of the direction of the replication fork movement.

Reciprocally inverse transformations of Clifford algebras $Cl(1, 2) \leftrightarrow Cl(0, 3)$ and $Cl(1, 2) \leftrightarrow Cl(2, 1)$ reveal the relativity of the leading chain choice and the possibility of circular DNA replication.

COMPLEMENTARY DNA PAIRS

Information about Clifford algebras *Cl*(2, 1) and *Cl*(3, 0), which are responsible for the beginning of synthesis of leading and lagging DNA chains, is contained in complementary DNA pairs.

Evidence of this statement is based on a mapping of hydrogen bonds of complementary DNA pairs (two typical and two possible ones, Figure 1.3) into metric tensor of Lorentz intervals.

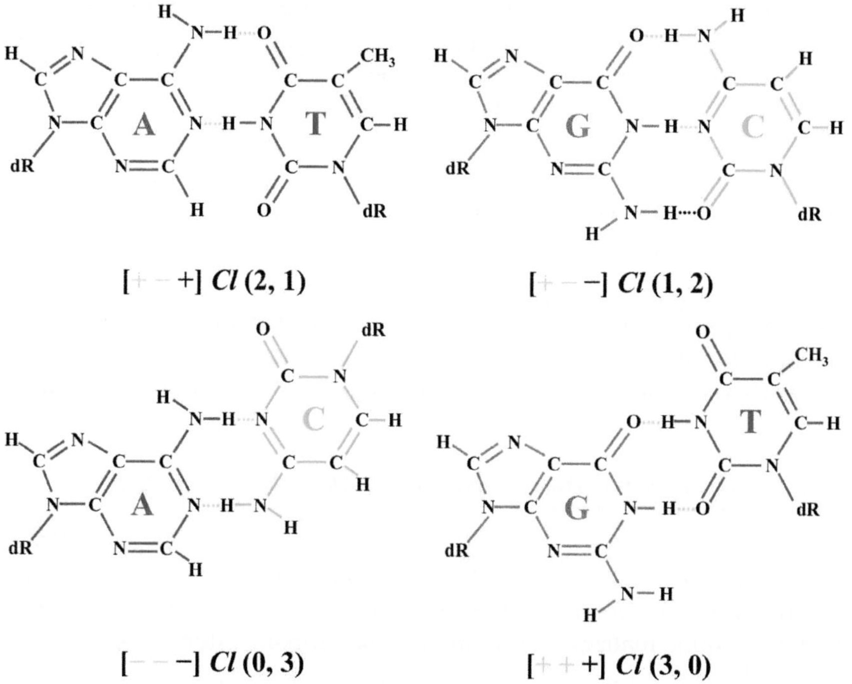

Figure 1.3 Clifford algebras of complementary DNA pairs. Intrinsic hydrogen bonds **NO** and **NN**, as well as signs of components of their metric tensor are shown in yellow and blue.

We supplement two common pairs **AT** and **GC** (Watson-Crick) with the pair **GT** (Wobble) and the reverse pair **AC** (Reverse wobble). These four complementary pairs are the map of four Clifford algebras *Cl*(3).

Let us consider features of the presented complementary pairs. First of all, the **GC** pair is distinguished by the availability of three hydrogen bonds. Suppose that excess bond **NO** in **GC** pair is connected with factor-metrics of hydrogen bonds of the complementary pairs.

Let us delete the excess (improper) bond **NO** in the **GC** pair. **AT** and **AC** pairs contain three proper hydrogen bonds **NN** and one proper hydrogen bond **NO** (metrics [1+ 3−]); **GC** and **GT** pairs contain one proper hydrogen bond **NN** and three proper hydrogen-bonds **NO** (metrics [3+ 1−]). Metrics may be chosen vice versa, but such choice would be erroneous.

Irrespective of chosen metrics, factor-metric of proper hydrogen bonds is equal [2+ 2−]. Sign of reset hydrogen bond **NO** in **GC** pair coincides with the sign of supplementary factor-metric in **AC** pair and is opposite to the signs of supplementary factor-metric in pairs **AT** and **GT**. This is because supplementary factor-metric should be orthogonal to the factor-metric of the proper hydrogen bonds. Coincidence of signs of the supplementary factor-metric (Euclidean metric) is erroneous. So it remains only to select the sign of the reset hydrogen bond **NO** in **GC** pair.

This sign may be only negative. All other variants of choosing signature signs in Clifford algebra, which correspond to hydrogen bonds of complementary DNA pairs, are erroneous, since they give a limited set of Clifford algebras $Cl(3)$ or lead to false algebra $Cl(2, 1)$ for the pair **GC**.

Clifford algebra $Cl(2, 1)$ of hydrogen bonds in the pair **AT** determines directly the origin of replication of the leading DNA chain, whereas Clifford algebra $Cl(1, 2)$ of the pair **GC** (which is isomorphic to the algebra $Cl(3, 0)$) indirectly determines the origin of replication of the lagging DNA chain.

CLIFFORD ALGEBRA OF DNA NUCLEOTIDES

Let us determine the signature (charges) of Clifford algebras of DNA nucleotides on the basis of vectors of donor-acceptor hydrogen bonds. Figure 1.4 shows possible vectors of hydrogen bonds of DNA nucleotides. Each nucleotide realizes presentation of the proper Clifford algebra:

$$\mathbf{T}Cl(2, 1), \mathbf{C}Cl(2, 2), \mathbf{A}Cl(2, 3), \mathbf{G}Cl(3, 3).$$

The number of vector-acceptors determines the positive part, whereas the number of vector-donors determines the negative part of the signature in Clifford algebra of nucleotides **C**, **T**, **G**, and vice versa—in Clifford algebra of nucleotide **A**. Why so?

To validate this statement, we should notice that algebra $Cl(0, 2)$ with generators in **AC** pair is Clifford algebra of the proper hydrogen bonds in the pairs **AT** and **AC**, whereas algebra $Cl(1, 1)$ with generators in **AC** pair is Clifford algebra of the proper hydrogen bonds in the pairs **GC** and **GT**.

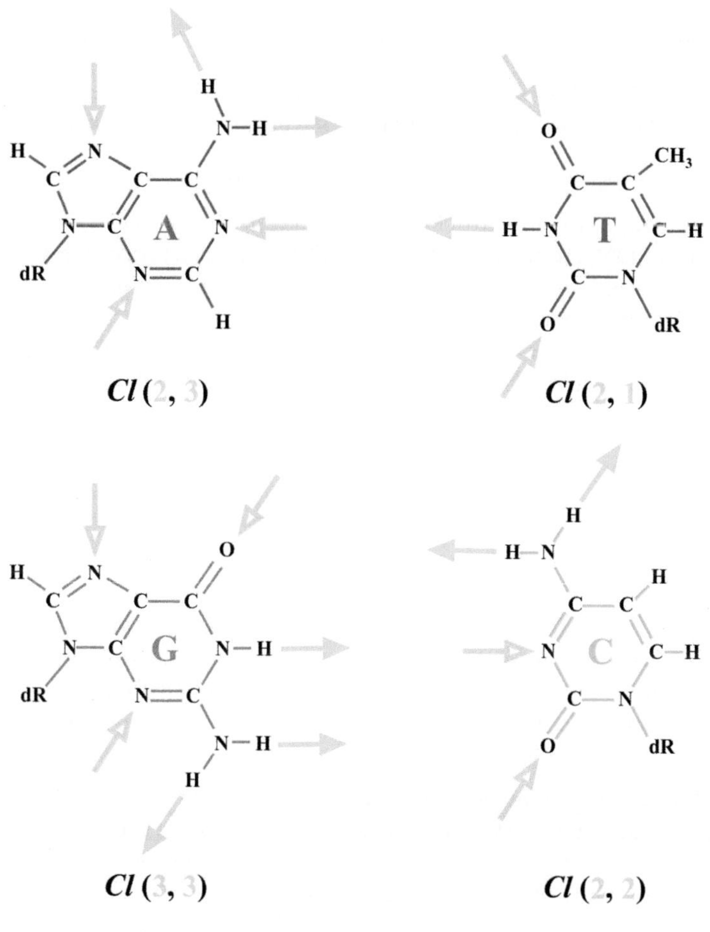

Figure 1.4 Potential vectors of hydrogen bond of DNA nucleotides. Yellow arrows—acceptors, blue arrows—donors of hydrogen.

Clifford algebras of proper hydrogen bonds in the complementary pairs **AT** and **GC** are factor-algebras of paired nucleotides:

$$\mathbf{A}Cl(2, 3) \cong \mathbf{T}Cl(2, 1) \cup Cl(0, 2); \mathbf{G}Cl(3, 3) \cong \mathbf{C}Cl(2, 2) \otimes Cl(1, 1).$$

At that, the following relationships are fulfilled:

$$\mathbf{A}Cl(2, 3) \otimes \mathbf{C}Cl(2, 2) \cong Cl(4, 5) = [256+ 256-],$$

$$\mathbf{T}Cl(2, 1) \otimes \mathbf{G}Cl(3, 3) \cong Cl(5, 4) = [272+ 240-].$$

DNA nucleotides have a different order of Clifford algebras; besides, these algebras are isomorphic to matrix algebras in the affine space:

$$\mathbf{A}Cl(2,3) \cong \text{Mat}(4, \mathbf{C}),\ \mathbf{C}Cl(2,2) \cong \text{Mat}(4, \mathbf{R}),$$
$$\mathbf{T}Cl(2,1) \cong \text{Mat}(2, \mathbf{R}) \oplus \text{Mat}(2, \mathbf{R}),\ \mathbf{G}Cl(3,3) \cong \text{Mat}(8, \mathbf{R}),$$

where **R, C** is the field of real and complex numbers.

Clifford algebra of the nucleotide **T(U)** coincides with Clifford algebra of the start of DNA replication. Usually **GT** pair does not occur in DNA; on the contrary, RNA contains the wobble pair **GU**. Numeric representation of DNA nucleotides on the basis of Clifford algebra shows that complementary DNA chains contain a mix of type I and II Lorentz intervals. To image DNA chains, it is enough to take only signs of squares of elements e_i, e_k belonging to Clifford algebras (DNA_{bin}):

5′3′-DNA: $A(+e_i^2)$ $T(-e_i^2)$ $C(+e_i^2)$ $G(-e_i^2)$
3′5′-DNA: $T(-e_k^2)$ $A(+e_k^2)$ $G(+e_k^2)$ $C(-e_k^2)$

The space of DNA nucleotide states contains

$$\mathbf{T}2^3 \otimes \mathbf{C}2^4 \otimes \mathbf{A}2^5 \otimes \mathbf{G}2^6 = 2^{18}$$

elements of Clifford algebras. This space reduction to four nucleotides means compression of DNA information by a factor of $2^{18}/4 = 65536$. Reduction of the nucleotide state space leads to DNA compactization and chromosome condensation.

Each DNA chain may be presented as a mix of Lorenz intervals from one Clifford algebra or as Lorenz intervals of the same type but from different Clifford algebras. Therefore, the "gene" may be interpreted as a mix of squares of elements of some Clifford algebras.

Compacted "gene" contains equal numbers of positive and negative squares; empty "gene" consists of **AT**-repeats. Clifford algebras of the type $Cl(p, p+1)$ always generate compacted "gene".

As an example, we will consider "gene" construction in Clifford algebra $Cl(9)$. Thereto, we will choose two algebras:

$$Cl(5,4),\ n_1 = 72,\ n_2 = 56;\ Cl(7,2),\ n_1 = 56,\ n_2 = 72,$$

where n_1 is the number of intervals [3+ 1–]; n_2—the number of intervals [1+ 3–].

5′3′ DNA chain contains 72 intervals of the algebra $Cl(5, 4)$ and 56 intervals of the algebra $Cl(7, 2)$; in all, 128 intervals [3+ 1–].

3′5′ DNA chain contains 72 intervals of the algebra $Cl(7, 2)$ and 56 intervals of the algebra $Cl(5, 4)$; totally, 128 intervals [1+ 3–].

Now we will compact the "gene" separating empty "gene" out of it (Figure 1.5).

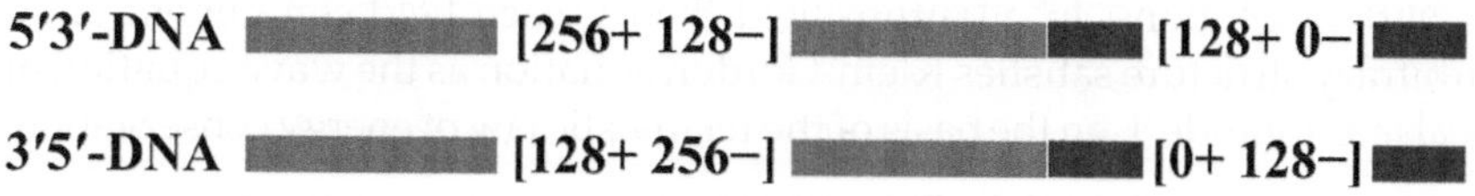

Figure 1.5 The structure of "gene" *Cl*(9). Compacted part of the "gene" is shown on the left, empty "gene" - on the right.

Figure 1.6 shows trial sequence of DNA nucleotides with Clifford algebras *Cl*(9). Further compactization of "gene" leads to increasing length of the empty "gene".

```
GAGTGCAAGGGAATCAAATAAAAGGACATAAAGCGAGACAGTGCGAAG
CTCACGTTCCCTTAGTTTATTTTCCTGTATTTCGCTCTGTCACGCTTC

CAGAAGAGCCCACCCCGCCCAAATCGAGCATGACGAAAAATAGCCACA
GTCTTCTCGGGTGGGGCGGGTTTAGCTCGTACTGCTTTTTATCGGTGT

ATCCCGGACACTGACGAACTAACACCTCCCTAGGAAACGAAGGACACG
TAGGGCCTGTGACTGCTTGATTGTGGAGGGATCCTTTGCTTCCTGTGC

CCCCAAGCAGGCAAGAGAAGACACGGAAGGGTCACCAATTAGGGAACA
GGGGTTCGTCCGTTCTCTTCTGTGCCTTCCCAGTGGTTAATCCCTTGT

AAAGAAGGTTAATGAAACTAAGGGAATAAGCACCGGACAAAATTACAA
TTTCTTCCAATTACTTTGATTCCCTTATTCGTGGCCTGTTTTAATGTT

ATACTGAGGCACACAAAACTGGGTCACGCAGAACTGCGCATAGAAGGA
TATGACTCCGTGTGTTTTGACCCAGTGCGTCTTGACGCGTATCTTCCT

AGACGCCACCCATGAACGGGGAATATAAAAAAAGCAGGCAAAATCAAT
TCTGCGGTGGGTACTTGCCCCTTATATTTTTTTCGTCCGTTTTAGTTA

ACGAGAAAAACACAAAAGCTTATTGACGGCAACCAAAGAAACGCGACC
TGCTCTTTTTGTGTTTTCGAATAACTGCCGTTGGTTTCTTTGCGCTGG

AAA...AAAAAAAAAAAAAAAAAAAAAAAAAAAAAAAAAAAAAAAAAA
TTT...TTTTTTTTTTTTTTTTTTTTTTTTTTTTTTTTTTTTTTTTTT
```

Figure 1.6 DNA nucleotide sequence of the "gene" *Cl*(9).

KLEIN-GORDON EQUATION

Let us consider the structure of DNA fragments on the basis of Klein-Gordon equation solutions. In quantum field theory, each field component of an arbitrary structure satisfies Klein-Gordon equation as the wave equation of scalar amplitude Ψ on the basis of the relativistic law of energy conservation:

$$(\mathbf{E}^2 - \mathbf{P}_x^{\ 2} - \mathbf{P}_y^{\ 2} - \mathbf{P}_z^{\ 2} - \mathbf{M}^2)\Psi = 0 \text{ (symbolic notation)},$$

where

$$\mathbf{E} = i\partial/\partial t,\ \mathbf{P}_x = -\,i\partial/\partial x,\ \mathbf{P}_y = -\,i\partial/\partial y,\ \mathbf{P}_z = -\,i\partial/\partial z$$

are operators of energy-impulse of massive (**M**) scalar field Ψ.

All equations will be written down in symbolic form.

Figure 1.7 shows generalized DNA structure.

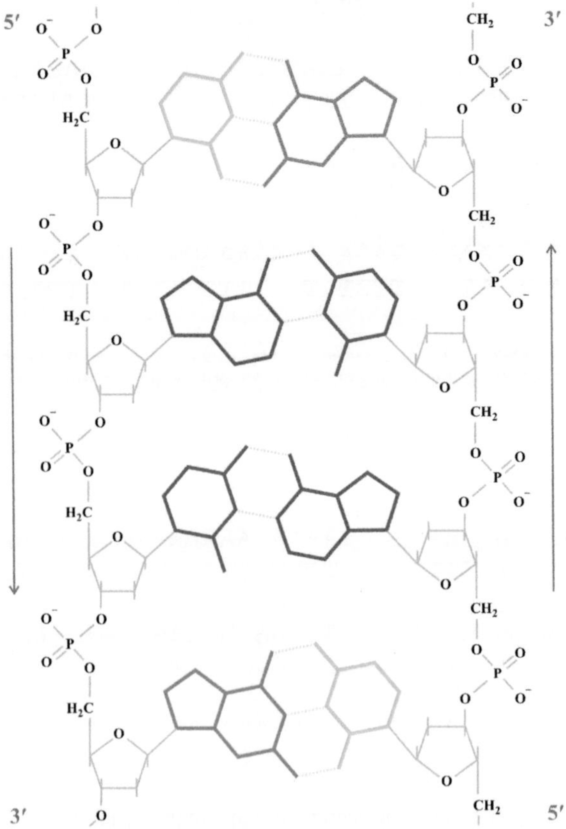

Figure 1.7 Generalized structure of DNA.

Synthesis of DNA structural equations is based on transformation of Klein-Gordon equation with utilization of Clifford algebras.

Let us direct **Z**-axis of three-dimensional right-hand reference systems in parallel to the helix axis of complementary DNA chains. Chains **5′3′** and **3′5′** are directed in antiparallel (supposedly, along **Z**-axis) and are determined by oppositely directed 5′ carbon atoms of 2-deoxyribose (**Z**-axis). Besides, in both chains, orientation of 2′ carbon atoms of 2-deoxyribose (**Y**-axis) is also mutually opposite, but the direction of **X**-axis of 2-deoxyribose coincides.

Let us turn **X**-axis in both chains in the opposite directions by the angle $\pi/4$. DNA nucleotide is linked up with **X**-axis of 2-deoxyribose. Nucleotides of complementary DNA pairs lie in **quantum mirror** (see Appendix 1.1).

Structural equations of 2-deoxyribose

In the Klein-Gordon equation, we will transfer operators $\mathbf{P}_x$ and $\mathbf{P}_y$ to the right (reflection in a plane parallel to **Z**-axis) and divide mass square into two parts:

$$(+\mathbf{E}^2 - \mathbf{P}_z^{\,2} + \mathbf{M}_1^{\,2})\Psi = (+\mathbf{P}_x^{\,2} + \mathbf{P}_y^{\,2} + \mathbf{M}_2^{\,2})\Psi;\ \mathbf{M}^2 = \mathbf{M}_2^{\,2} - \mathbf{M}_1^{\,2}.$$

Suppose that operators **E** and $\mathbf{P}_z$ are generators of Clifford algebra $Cl(1, 1)$, and operators $\mathbf{P}_x$ and $\mathbf{P}_y$ are generators of Clifford algebra $Cl(2, 0)$; therefore, structural equation of 2-deoxyribose is:

$$\{Cl(1, 1) + \mathbf{M}_1^{\,2}\}\Psi = \{Cl(2, 0) + \mathbf{M}_2^{\,2}\}\Psi$$

Or, subject to equation $Cl(1, 1) \cong Cl(2, 0) = [3+\ 1-]$, we have:

$$(\mathbf{E}_1^{\,2} - \mathbf{P}_{x1}^{\,2} - \mathbf{P}_{y1}^{\,2} - \mathbf{P}_{z1}^{\,2} - \mathbf{M}_1^{\,2})\Psi = (\mathbf{E}_2^{\,2} - \mathbf{P}_{x2}^{\,2} - \mathbf{P}_{y2}^{\,2} - \mathbf{P}_{z2}^{\,2} - \mathbf{M}_2^{\,2})\Psi;$$

$$(\mathbf{E}_{21}^{\,2} - \mathbf{P}_{x21}^{\,2} - \mathbf{P}_{y21}^{\,2} - \mathbf{P}_{z21}^{\,2} - \mathbf{M}^2)\Psi_{21} = 0;\ \mathbf{E}_{21}^{\,2} = \mathbf{E}_2^{\,2} - \mathbf{E}_1^{\,2},$$

$$\mathbf{P}_{x21}^{\,2} = \mathbf{P}_{x2}^{\,2} - \mathbf{P}_{x1}^{\,2},\ \mathbf{P}_{y21}^{\,2} = \mathbf{P}_{y2}^{\,2} - \mathbf{P}_{y1}^{\,2},\ \mathbf{P}_{z21}^{\,2} = \mathbf{P}_{z2}^{\,2} - \mathbf{P}_{z1}^{\,2}.$$

Ψ_{21} field is a difference field of complementary DNA chains or of the pair: DNA chain—RNA chain.

Pairs 2-deoxyribose (Ψ_1)—2-deoxyribose (Ψ_2) or 2-deoxyribose (Ψ_1)—ribose (Ψ_2) satisfy the structural equations:

$$(\mathbf{E}_1^{\,2} - \mathbf{P}_{x1}^{\,2} - \mathbf{P}_{y1}^{\,2} - \mathbf{P}_{z1}^{\,2} - \mathbf{M}_1^{\,2})\Psi_1 = 0;$$

$$(\mathbf{E}_2^{\,2} - \mathbf{P}_{x2}^{\,2} - \mathbf{P}_{y2}^{\,2} - \mathbf{P}_{z2}^{\,2} - \mathbf{M}_2^{\,2})\Psi_2 = 0.$$

Field functions Ψ_1, Ψ_2, Ψ_{21} form two interpenetrating coaxial right-handed helices with different masses. The relation $\mathbf{M}_1 \neq \mathbf{M}_2$ is always correct, since masses of complementary nucleotides, associated with 2-deoxyribose (ribose), are different.

If $\mathbf{M}_1 = \mathbf{M}_2$, then the field functions Ψ_1, Ψ_2 would form massive complex scalar field $\Psi = \Psi_1 \pm i\Psi_2$, and massless field Ψ_{21} would determine two filaments of right-hand helix.

Structural equations of phosphates

Phosphate fragments of complementary DNA chains are mutually turned by angle π about **X** axis, which is perpendicular to the axis of DNA helix, therefore, two phosphate fragments form a two-component scalar field.

Let us record the equation of Klein-Gordon for phosphates in the form:

$$(-\mathbf{E}^2 + \mathbf{P}_z^{\,2} + \mathbf{M}_1^{\,2})\Psi = (-\mathbf{P}_x^{\,2} - \mathbf{P}_y^{\,2} - \mathbf{M}_2^{\,2})\Psi;\ \mathbf{M}^2 = \mathbf{M}_1^{\,2} + \mathbf{M}_2^{\,2},\ \text{(KGph)}$$

square of mass **M** in the equation is represented as the sum of two squares of phosphate masses. Phosphate masses may differ in the mass of dissociated hydrogen, but, on the whole, $\mathbf{M}_1 = \mathbf{M}_2$.

Let us transform Klein-Gordon equation, supposing that the operators **E** and $\mathbf{P}_z$ are generators of Clifford algebra $Cl(1, 1)$, and the operators $\mathbf{P}_x$ and $\mathbf{P}_y$ are generators of Clifford algebra $Cl(0, 2)$:

$$\{Cl(1, 1) + \mathbf{M}_1^{\,2}\}\Psi = \{Cl(0, 2) - \mathbf{M}_2^{\,2}\}\Psi.$$

We will have the following structural equations of phosphates:

$$-(\mathbf{E}_1^{\,2} - \mathbf{P}_{x1}^{\,2} - \mathbf{P}_{y1}^{\,2} - \mathbf{P}_{z1}^{\,2} - \mathbf{M}_1^{\,2})\Psi = (\mathbf{E}_2^{\,2} - \mathbf{P}_{x2}^{\,2} - \mathbf{P}_{y2}^{\,2} - \mathbf{P}_{z2}^{\,2} - \mathbf{M}_2^{\,2})\Psi,$$

$$(\mathbf{E}_{12}^{\,2} - \mathbf{P}_{x12}^{\,2} - \mathbf{P}_{y12}^{\,2} - \mathbf{P}_{z12}^{\,2} - \mathbf{M}^2)\Psi_{12} = 0;\ \mathbf{E}_{12}^{\,2} = \mathbf{E}_1^{\,2} + \mathbf{E}_2^{\,2},$$

$$\mathbf{P}_{x12}^{\,2} = \mathbf{P}_{x1}^{\,2} + \mathbf{P}_{x2}^{\,2},\ \mathbf{P}_{y12}^{\,2} = \mathbf{P}_{y1}^{\,2} + \mathbf{P}_{y2}^{\,2},\ \mathbf{P}_{z12}^{\,2} = \mathbf{P}_{z1}^{\,2} + \mathbf{P}_{z2}^{\,2},$$

$$(\mathbf{E}_1^{\,2} - \mathbf{P}_{x1}^{\,2} - \mathbf{P}_{y1}^{\,2} - \mathbf{P}_{z1}^{\,2} - \mathbf{M}_1^{\,2})\Psi_1 = 0,$$

$$(\mathbf{E}_2^{\,2} - \mathbf{P}_{x2}^{\,2} - \mathbf{P}_{y2}^{\,2} - \mathbf{P}_{z2}^{\,2} - \mathbf{M}_2^{\,2})\Psi_2 = 0.$$

Average summary field Ψ_{12} is a two-component field, which may satisfy gauge transformations, that is, rotations in **XY** plane.

Structural equations of complementary pairs

Let us invert left and right parts of the equation (KGph):

$$(+\mathbf{E}^2 - \mathbf{P}_z^{\,2} - \mathbf{M}_1^{\,2})\Psi = (+\mathbf{P}_x^{\,2} + \mathbf{P}_y^{\,2} + \mathbf{M}_2^{\,2})\Psi;\ \mathbf{M}^2 = \mathbf{M}_1^{\,2} + \mathbf{M}_2^{\,2}.$$

Suppose that operators $\mathbf{E}$, $\mathbf{P}_z$, $\mathbf{M}_1$ are generators of Clifford algebra $Cl(1, 2)$, and operators $\mathbf{P}_x$, $\mathbf{P}_y$, $\mathbf{M}_2$ are generators of Clifford algebra $Cl(3, 0)$. We will get equivalence of Clifford algebras:

$$Cl(1,2)\Psi = Cl(3,0)\Psi;\ Cl(1,2) \cong Cl(3,0) = [1+\ 3-] \oplus [3+\ 1-].$$

Let us write down the equation of complementary DNA pairs with regard to nucleotide masses:

$$(\mathbf{E}_1^{\,2} - \mathbf{P}_{x1}^{\,2} - \mathbf{P}_{y1}^{\,2} - \mathbf{P}_{z1}^{\,2} - \mathbf{M}_{01}^{\,2})\Psi - (\mathbf{E}_2^{\,2} - \mathbf{P}_{x2}^{\,2} - \mathbf{P}_{y2}^{\,2} - \mathbf{P}_{z2}^{\,2} - \mathbf{M}_{02}^{\,2})\Psi =$$
$$= (\mathbf{E}_3^{\,2} - \mathbf{P}_{x3}^{\,2} - \mathbf{P}_{y3}^{\,2} - \mathbf{P}_{z3}^{\,2} - \mathbf{M}_{03}^{\,2})\Psi - (\mathbf{E}_4^{\,2} - \mathbf{P}_{x4}^{\,2} - \mathbf{P}_{y4}^{\,2} - \mathbf{P}_{z4}^{\,2} - \mathbf{M}_{04}^{\,2})\Psi.$$

Suppose that all masses in brackets are equal to zero, then all four components of Ψ field should satisfy massless (that is, informational) Klein-Gordon equation:

$$(\mathbf{E}_{1,2,3,4}^{\,2} - \mathbf{P}_{x1,2,3,4}^{\,2} - \mathbf{P}_{y1,2,3,4}^{\,2} - \mathbf{P}_{z1,2,3,4}^{\,2})\Psi_{1,2,3,4} = 0.$$

Informational DNA nucleotides are massive molecules; therefore, the following structural equation is valid for nucleotides:

$$(\mathbf{E}_{1,2,3,4}^{\,2} - \mathbf{P}_{x1,2,3,4}^{\,2} - \mathbf{P}_{y1,2,3,4}^{\,2} - \mathbf{P}_{z1,2,3,4}^{\,2} - \mathbf{M}_{01,2,3,4}^{\,2})\Psi_{1,2,3,4} = 0.$$

Informational nucleotides will be massive, if the condition of balance of field nucleotide masses is satisfied:

$$-\mathbf{M}_{01}^{\,2} + \mathbf{M}_{02}^{\,2} = -\mathbf{M}_{03}^{\,2} + \mathbf{M}_{04}^{\,2}.$$

Let us rearrange brackets of the structural equation of complementary DNA pairs:

$$(\mathbf{E}_1^{\,2} - \mathbf{P}_{x1}^{\,2} - \mathbf{P}_{y1}^{\,2} - \mathbf{P}_{z1}^{\,2} - \mathbf{M}_{01}^{\,2})\Psi + (\mathbf{E}_4^{\,2} - \mathbf{P}_{x4}^{\,2} - \mathbf{P}_{y4}^{\,2} - \mathbf{P}_{z4}^{\,2} - \mathbf{M}_{04}^{\,2})\Psi =$$
$$= (\mathbf{E}_2^{\,2} - \mathbf{P}_{x2}^{\,2} - \mathbf{P}_{y2}^{\,2} - \mathbf{P}_{z2}^{\,2} - \mathbf{M}_{02}^{\,2})\Psi + (\mathbf{E}_3^{\,2} - \mathbf{P}_{x3}^{\,2} - \mathbf{P}_{y3}^{\,2} - \mathbf{P}_{z3}^{\,2} - \mathbf{M}_{03}^{\,2})\Psi.$$

$$(\mathbf{E}_1^{\,2} - \mathbf{P}_{x1}^{\,2} - \mathbf{P}_{y1}^{\,2} - \mathbf{P}_{z1}^{\,2} - \mathbf{M}_{01}^{\,2})\Psi_{1,4} + (\mathbf{E}_4^{\,2} - \mathbf{P}_{x4}^{\,2} - \mathbf{P}_{y4}^{\,2} - \mathbf{P}_{z4}^{\,2} - \mathbf{M}_{04}^{\,2})\Psi_{1,4} = 0,$$
$$(\mathbf{E}_2^{\,2} - \mathbf{P}_{x2}^{\,2} - \mathbf{P}_{y2}^{\,2} - \mathbf{P}_{z2}^{\,2} - \mathbf{M}_{02}^{\,2})\Psi_{2,3} + (\mathbf{E}_3^{\,2} - \mathbf{P}_{x3}^{\,2} - \mathbf{P}_{y3}^{\,2} - \mathbf{P}_{z3}^{\,2} - \mathbf{M}_{03}^{\,2})\Psi_{2,3} = 0.$$
$$\mathbf{M}_{01}^{\,2} + \mathbf{M}_{04}^{\,2} = \mathbf{M}_{02}^{\,2} + \mathbf{M}_{03}^{\,2}.$$

For complementary DNA pairs, mass relation M(**TA**) ≈ M(**CG**) is performed:

T =	5C	5H	2O	2N	=	65*p*	60*n*
A =	5C	4H		5N	=	69*p*	65*n*
				TA	=	134*p*	125*n*
C =	4C	4H	1O	3N	=	57*p*	53*n*
G =	5C	4H	1O	5N	=	77*p*	73*n*
				CG	=	134*p*	126*n*

Here the structure of nucleotides is represented by the number of atoms **C, H, O, N** and by the number of protons p and neutrons n. As compared with **TA** pair, the **CG** pair contains one excessive neutron. This neutron belongs to nitrogen atom **N** in position 7 of **G** nucleotide and is connected with factor-metric [3+ 1–] of 7-methylguanine of eukaryote mRNA cap structure (Chapter 3). Excessive neutron of 7-methylguanine and three hydrogen atoms of methyl group $\mathbf{CH_3}$ form metric of mRNA translation intervals: $[\mathbf{H}_3\, n] = [3+\ 1-]$.

Excessive neutron in **CG** pair may be compensated by the addition of an hydrogen atom with formation of hydroxyl in one of the **TA** pair phosphates. Therefore, one may suppose that

$$\mathrm{M}(\mathbf{T}) + \mathrm{M}(\mathbf{A}) = \mathrm{M}(\mathbf{C}) + \mathrm{M}(\mathbf{G}).$$

Complementary pair fields will be chosen as follows:

$$\Psi_{1,4} = \Psi(\mathbf{TA}),\ \Psi_{2,3} = \Psi(\mathbf{CG});$$

$$\mathbf{M}_{01}{}^2 = \mathrm{M}_0\mathrm{M}(\mathbf{T}),\ \mathbf{M}_{04}{}^2 = \mathrm{M}_0\mathrm{M}(\mathbf{A}),\ \mathbf{M}_{02}{}^2 = \mathrm{M}_0\mathrm{M}(\mathbf{C}),\ \mathbf{M}_{03}{}^2 = \mathrm{M}_0\mathrm{M}(\mathbf{G}),$$

where M_0 is an integrating mass of complementary nucleotides.

Structural equations of nucleotide triplets

Let us redesignate mass operators in the equation (KGph):

$$(-\mathbf{E}^2 + \mathbf{P}_z{}^2 + \mathbf{M}_1{}^2)\Psi = (-\mathbf{P}_x{}^2 - \mathbf{P}_y{}^2 - \mathbf{M}_3{}^2)\Psi;\ \mathbf{M}^2 = \mathbf{M}_1{}^2 + \mathbf{M}_3{}^2.$$

With the assumption that operators $\mathbf{E}$, $\mathbf{P}_z$, $\mathbf{M}_1$ are generators of Clifford algebra $Cl(2, 1)$, and operators $\mathbf{P}_x$ and $\mathbf{P}_y$ are generators of Clifford algebra $Cl(0, 2)$ we will have:

$$Cl(2, 1)\Psi = \{Cl(0, 2) - \mathbf{M}_3{}^2\}\Psi.$$

We have structural equations of nucleotide triplets:

$$-(\mathbf{E}_1{}^2 - \mathbf{P}_{x1}{}^2 - \mathbf{P}_{y1}{}^2 - \mathbf{P}_{z1}{}^2)\Psi - (\mathbf{E}_2{}^2 - \mathbf{P}_{x2}{}^2 - \mathbf{P}_{y2}{}^2 - \mathbf{P}_{z2}{}^2)\Psi =$$
$$= (\mathbf{E}_3{}^2 - \mathbf{P}_{x3}{}^2 - \mathbf{P}_{y3}{}^2 - \mathbf{P}_{z3}{}^2 - \mathbf{M}_3{}^2)\Psi$$

or

$$(\mathbf{E}_1{}^2 - \mathbf{P}_{x1}{}^2 - \mathbf{P}_{y1}{}^2 - \mathbf{P}_{z1}{}^2)\Psi + (\mathbf{E}_2{}^2 - \mathbf{P}_{x2}{}^2 - \mathbf{P}_{y2}{}^2 - \mathbf{P}_{z2}{}^2)\Psi +$$
$$+ (\mathbf{E}_3{}^2 - \mathbf{P}_{x3}{}^2 - \mathbf{P}_{y3}{}^2 - \mathbf{P}_{z3}{}^2 - \mathbf{M}_3{}^2)\Psi = 0.$$

Ψ field is a three-component field with mass, which is equal to the nucleotide mass in the third position of triplet; first and second nucleotides do not make a contribution to the field triplet mass. This circumstance may be a cause of genetic code degeneracy.

APPENDIX 1.1

MATHEMATICS OF CLIFFORD ALGEBRA

Clifford algebra was developed by English mathematician William Clifford in 1878.

Let E be a vector space over field of real numbers **R** or complex numbers **C**. Dimension of E space is dim = 2^n, where n is a natural number.

Let us introduce basis {V} in E from 2^n elements with naturally ordered indices $a_1 < a_2 < \ldots$ from 1 to n:

$$e_0, e_a, e_{a1a2}, e_{a1a2a3}, e_{a1a2a3\ldots}, e_{1\ldots n},$$

which is numbered by multi indices with length from 1 to n.

Let $p + q = n; p > 0, q > 0$. Let us introduce the diagonal matrix $||\eta_{ab}||$ with dimension $n \times n$:

$$||\eta_{ab}|| = \mathrm{diag}(+1, \ldots, +1, -1, \ldots, -1),$$

diagonal of the matrix contains p positive units and q negative units.

Clifford algebra is a type of associative algebra with the following relationships:

1) e_0 is the unity of algebra, commutes with all basic elements {V} of the algebra

$$e_0\{V\} = \{V\}e_0 = \{V\};$$

2) for all indices $a, b \in 1, 2, \ldots, n$, the generators (**charges**) anticommute between themselves

$$e_a e_b + e_b e_a = 2\eta_{ab};$$

3) for all indices $a_1 \leq a_2 \ldots \leq a_k$ from 1 to n:

$$e_{a1} \ldots e_{ak} = e_{a1\ldots ak}.$$

Number pair (p, q) is called main signature or, simply, signature of Clifford algebra $Cl(p, q)$. Exact position of elements +1 and –1 on diagonal of the matrix $||\eta_{ab}||$ is called sign signature. All Clifford algebras with different sign signatures are isomorphic between themselves, but subsets of basic elements may be nonisomorphic. Clifford algebra with $p + q = n$ is designated as $Cl(n)$.

At calculating squares of basic elements of Clifford algebra, key role is assigned to ratio of revers:

$$e_{a1\ldots ak} = (-1)^{k(k-1)/2}\, e_{ak\ldots a1}.$$

Below, we present values of revers factor for different values of k index:

Index k	0	1	2	3	4	5	6	7	8	9
$(-1)^{k(k-1)/2}$	+1	+1	−1	−1	+1	+1	−1	−1	+1	+1

Clifford algebras with small dimensions are isomorphic to well-known algebras:

$$Cl(0,0) = \mathbf{R},\ Cl(1,0) = \mathbf{R} \oplus \mathbf{R},\ Cl(0,1) = \mathbf{C}.$$
$$Cl(1,1) = Cl(2,0) = \mathrm{Mat}(2, \mathbf{R}),\ Cl(0,2) = \mathbf{H},$$

where

Mat(2, **R**) is matrix algebra of second order in the field of real numbers;

H is algebra of quaternions.

In Clifford algebra, the following relations are performed:

$$\boldsymbol{Cl\,(p, q) \cong Cl\,(q+1, p-1);\ Cl\,(p, q) \cong Cl\,(p-4, q+4);}$$
$$\boldsymbol{Cl\,(p+1, q+1) \cong Cl\,(p, q) \otimes Cl\,(1, 1);}$$
$$\boldsymbol{Cl\,(p, p+1) \oplus Cl\,(p+1, p) \sim Cl\,(p+1, p+1);}$$
$$\boldsymbol{Cl\,(p, p+1) = [S+\ S-],}$$

where **[P+ Q–]** is a complete signature of all basic elements of Clifford algebra.

Clifford algebra $Cl(p, q)$ (where $p + q = n$) is equivalent to matrix algebra:

$$Cl(p,q) = \begin{cases} \mathrm{Mat}(2^{n/2}, \mathbf{R}), & \text{if } p-q = 0, 2 \bmod 8; \\ \mathrm{Mat}(2^{(n-1)/2}, \mathbf{R}) \oplus \mathrm{Mat}(2^{(n-1)/2}, \mathbf{R}), & \text{if } p-q = 1 \bmod 8; \\ \mathrm{Mat}(2^{(n-1)/2}, \mathbf{C}), & \text{if } p-q = 3, 7 \bmod 8; \\ \mathrm{Mat}(2^{(n-2)/2}, \mathbf{H}), & \text{if } p-q = 4, 6 \bmod 8; \\ \mathrm{Mat}(2^{(n-3)/2}, \mathbf{H}) \oplus \mathrm{Mat}(2^{(n-3)/2}, \mathbf{H}), & \text{if } p-q = 5 \bmod 8. \end{cases}$$

Complete signature **[P+ Q–]** (where **(P+) + (Q–)** $= 2^n$) of Clifford algebra $Cl(p, q)$ may be presented as n_1 Lorentz intervals [3+ 1–] and n_2 Lorentz intervals [1+ 3–], where

$$n_1 = \{3(\mathbf{P+}) - (\mathbf{Q-})\}/8;\ n_2 = \{3(\mathbf{Q-}) - (\mathbf{P+})\}/8.$$

With the rise of Clifford algebra order, numbers n_1 and n_2 approach each other:

$$Cl(6,4) = [528+\ 496-],\ n_1 = 136,\ n_2 = 120;$$
$$Cl(7,7) = [8256+\ 8128-],\ n_1 = 2080,\ n_2 = 2016;$$
$$Cl(10,10) = [524800+\ 523776-],\ n_1 = 131328,\ n_2 = 130816.$$

Every Lorentz interval [3+ 1–] may be transferred into Lorentz interval [1+ 3–] through replacement of real basis by the imaginary one and vice versa, that is, both intervals are a reflection of a each other in quantum mirror.

Quantum mirror is the reflection of physical object in four or three planes at the angle $\pi/2$:

$$x \to -y, y \to -z, z \to -t, t \to -x,$$

right-handed screw → right-handed screw;

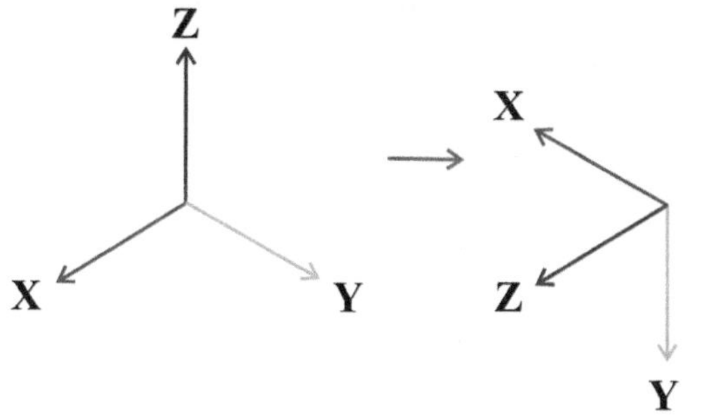

or

$$x \to -y, y \to -z, z \to -x,$$

right-handed screw → left-handed screw.

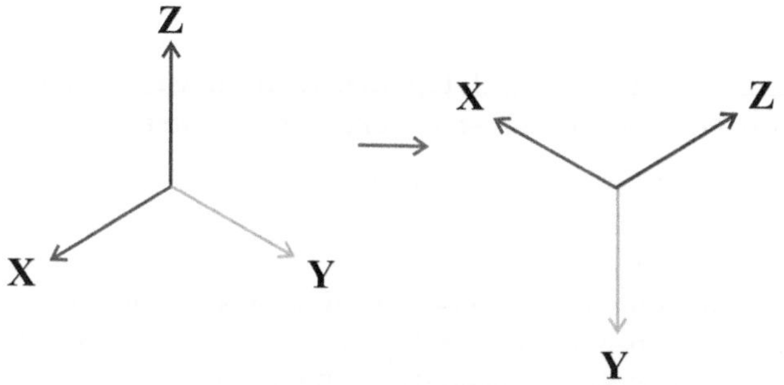

If two spirals at a reflection in a quantum mirror are directed in opposite directions, and identically screwed up in the three-dimensional world, then they are directed and screwed up in the four-dimensional world in opposite directions. This probably plays a role at separation of DNA chains.

We will separate Lorentz intervals of I and II type into discrete sets Ω_I = {1+ 3–} и Ω_{II} = {3+ 1–}. Each Lorentz interval is a point of four-dimensional space (real or imaginary). Let us connect points of the sets Ω_I and Ω_{II} with separate chains. If numbers n_1 and n_2 are not equal, chain of real intervals will not coincide with the chain of imaginary intervals.

Chain of real Lorentz intervals will change its length until it coincides with the chain length of imaginary intervals—such is the nontrivial conclusion of non-commutative geometry.

Let $n = 2m$. Then it is not difficult to prove: if **(P+)** = 0 mod 3, then **(Q–)** = 1 mod 3 and vice versa.

Let us consider two cases.

1) **(P+)** = 1 mod 3, **(Q–)** = 0 mod 3. Associated with time variable $t(e_0)$ from the set **(P+)** together with three spatial variables from the set **(Q–)** form time-like Lorentz interval

$$[1+3-] = t^2(e_0) - x^2(e_\alpha) - y^2(e_\beta) - z^2(e_\gamma)$$

of imaging point.

We will allocate the set $\{$**(P+) – 1**$\}$ by three-dimensional coordinates of $\{$**(P+) – 1**$\}/3$ points and connect them into a chain. The set $\{$**(Q–) – 3**$\}$ will also be distributed by three-dimensional coordinates of $\{$**(Q–) – 3**$\}/3$ points with their connection into a chain.

Imaging point shows that chain of $\{$**(Q–) – 3**$\}/3$ points remains in a local region of real three-dimensional space, and the chain, including $\{$**(P+) – 1**$\}/3$ points, is located in the mirror. Chain of $\{$**(Q–) – 3**$\}/3$ points moves so that its length will be equal to the length of motionless chain of $\{$**(P+) – 1**$\}/3$ points located in the mirror.

2) **(P+)** = 0 mod 3, **(Q–)** = 1 mod 3. Variable $t(e_\delta)$, associated with time, is chosen from the set **(Q–); ** together with three space coordinates of the set **(P+)**, it determines space-like Lorentz interval

$$[3+1-] = x^2(e_0) + y^2(e_v) + z^2(e_w) - t^2(e_\delta)$$

of the imaging point.

We will distribute the set $\{$**(Q–) – 1**$\}$ by three-dimensional coordinates of $\{$**(Q–) – 1**$\}/3$ points and connect them into a chain. The set $\{$**(P+) – 3**$\}$ will be distributed by three-dimensional coordinates of $\{$**(P+) – 3**$\}/3$ points and also connected into a chain.

Imaging point shows that the chain of $\{$**(P+) – 3**$\}/3$ points is located in some region of real three-dimensional space, and the chain of $\{$**(Q–) – 1**$\}/3$ points is located in the mirror. Chain points of the set $\{$**(P+) – 3**$\}/3$ move synchronously so that the length of their chain will be equal to the length of immovable chain of $\{$**(Q–) – 1**$\}/3$ points allocated in the mirror.

In the first case, the time commutates with the coordinates of the imaging point, which corresponds to **reversible** change of the chain of {**(Q–) – 1**}/3 points.

In the second case, the time commutates only with one coordinate $x(e_0)$ of imaging point, which corresponds to **nonreversible** change of the chain of {**(P+) – 3**}/3 points.

APPENDIX 1.2

HADAMARD MATRICES OF CLIFFORD ALGEBRAS

Clifford algebras of the same order but with different sign signature produce Hadamard matrix.

Let us arrange horizontally square values of all elements of Clifford algebra $Cl(n)$ in natural order, beginning from e_0^2 and finishing by $e_{1...n}^2$. In each row of matrix $2^n \times 2^n$ we will change sign signature of Clifford algebra. We will have Hadamard matrix of 2^n order.

The first and last rows of Hadamard matrix $Cl(n)$ contain n noncoincidences between generators, and all noncoincidences of row elements are located in sections of C^n_k elements odd by k, where C^n_k are binomial coefficients.

In general, product of elements of two rows in Hadamard matrix $Cl(n)$ contains p coincidences and q noncoincidences between generators $(p + q = n)$. But now noncoincidences of elements in two rows appears also in sections of C^n_k elements even by k.

The sum of all noncoincidences of elements in two rows is equal to 2^{n-1} that follows from the identity

$$\sum_{k = 1...n;\, m = 1,3,5...} \sum C^q_m C^p_{k-m} = 2^{p+q-1}$$

which proves orthogonality of Hadamard matrix $Cl(n)$.

Figure 1.8 shows Hadamard matrices of low order Clifford algebras. We use the following notations: "1" = +1, "–" = –1.

***Cl*(1)**

1	-
1	1

***Cl*(2)**

1	-	-	-
1	1	-	1
1	-	1	1
1	1	1	-

***Cl*(3)**

1	-	-	-	-	-	-	1
1	1	-	-	1	1	-	-
1	-	1	-	1	-	1	-
1	-	-	1	-	1	1	-
1	1	1	-	-	1	1	1
1	1	-	1	1	-	1	1
1	-	1	1	1	1	-	1
1	1	1	1	-	-	-	-

***Cl*(4)**

1	-	-	-	-	-	-	-	-	-	-	1	1	1	1	1
1	1	-	-	-	1	1	1	-	-	-	-	-	-	1	-
1	-	1	-	-	1	-	-	1	1	-	-	-	1	-	-
1	-	-	1	-	-	1	-	1	-	1	-	1	-	-	-
1	-	-	-	1	-	-	1	-	1	1	1	-	-	-	-
1	1	1	-	-	-	1	1	1	1	-	1	1	-	-	1
1	1	-	1	-	1	-	1	1	-	1	1	-	1	-	1
1	1	-	-	1	1	1	-	-	1	1	-	1	1	-	1
1	-	1	1	-	1	1	-	-	1	1	1	-	-	1	1
1	-	1	-	1	1	-	1	1	-	1	-	1	-	1	1
1	-	-	1	1	-	1	1	1	1	-	-	-	1	1	1
1	1	1	1	-	-	-	1	-	1	1	-	1	1	1	-
1	1	1	-	1	-	1	-	1	-	1	1	-	1	1	-
1	1	-	1	1	1	-	-	1	1	-	1	1	-	1	-
1	-	1	1	1	1	1	1	-	-	-	1	1	1	-	-
1	1	1	1	1	-	-	-	-	-	-	-	-	-	-	1

***Cl*(5)**

1	-	-	-	-	-	-	-	-	-	-	-	-	-	-	-	1	1	1	1	1	1	1	1	1	1	1	1	1	1	1	-
1	1	-	-	-	-	1	1	1	1	-	-	-	-	-	-	-	-	-	-	-	-	1	1	1	1	-	-	-	-	1	1
1	-	1	-	-	-	1	-	-	-	1	1	1	-	-	-	-	-	-	1	1	1	-	-	-	1	-	-	-	1	-	1
1	-	-	1	-	-	-	1	-	-	1	-	-	1	1	-	-	1	1	-	-	1	-	-	1	-	-	-	1	-	-	1
1	-	-	-	1	-	-	-	1	-	-	1	-	1	-	1	1	-	1	-	1	-	-	1	-	-	-	1	-	-	-	1
1	-	-	-	-	1	-	-	-	1	-	-	1	-	1	1	1	1	-	1	-	-	1	-	-	-	1	-	-	-	-	1
1	1	1	-	-	-	-	1	1	1	1	1	1	-	-	-	1	1	1	-	-	-	-	-	-	1	1	1	1	-	-	-
1	1	-	1	-	-	1	-	1	1	1	-	-	1	1	-	1	-	-	1	1	-	-	-	1	-	1	1	-	1	-	-
1	1	-	-	1	-	1	1	-	1	-	1	-	1	-	1	-	1	-	1	-	1	-	1	-	-	1	-	1	1	-	-
1	1	-	-	-	1	1	1	1	-	-	-	1	-	1	1	-	-	1	-	1	1	1	-	-	-	-	1	1	1	-	-
1	-	1	1	-	-	1	1	-	-	-	1	1	1	1	-	1	-	-	-	-	1	1	1	-	-	1	1	-	-	1	-
1	-	1	-	1	-	1	-	1	-	1	-	1	1	-	1	-	1	-	-	1	-	1	-	1	-	1	-	1	-	1	-
1	-	1	-	-	1	1	-	-	1	1	1	-	-	1	1	-	-	1	1	-	-	-	1	1	-	-	1	1	-	1	-
1	-	-	1	1	-	-	1	1	-	1	1	-	-	1	1	-	-	1	1	-	-	1	-	-	1	1	-	-	1	1	-
1	-	-	1	-	1	-	1	-	1	1	-	1	1	-	1	-	1	-	-	1	-	-	1	-	1	-	1	-	1	1	-
1	-	-	-	1	1	-	-	1	1	-	1	1	1	1	-	1	-	-	-	-	1	-	-	1	1	-	-	1	1	1	-
1	1	1	1	-	-	-	-	1	1	-	1	1	1	1	-	-	1	1	1	1	-	1	1	-	-	-	-	1	1	1	1
1	1	1	-	1	-	-	1	-	1	1	-	1	1	-	1	1	-	1	1	-	1	1	-	1	-	-	1	-	1	1	1
1	1	1	-	-	1	-	1	1	-	1	1	-	-	1	1	1	1	-	-	1	1	-	1	1	-	1	-	-	1	1	1
1	1	-	1	1	-	1	-	-	1	1	1	-	-	1	1	1	1	-	-	1	1	1	-	-	1	-	1	1	-	1	1
1	1	-	1	-	1	1	-	1	-	1	-	1	1	-	1	1	-	1	1	-	1	-	1	-	1	1	-	1	-	1	1
1	1	-	-	1	1	1	1	-	-	-	1	1	1	1	-	-	1	1	1	1	-	-	-	1	1	1	1	-	-	1	1
1	-	1	1	1	-	1	1	1	-	-	-	1	-	1	1	1	1	-	1	-	-	-	1	1	1	-	1	1	1	-	1
1	-	1	1	-	1	1	1	-	1	-	1	-	1	-	1	1	-	1	-	1	-	1	-	1	1	1	-	1	1	-	1
1	-	1	-	1	1	1	-	1	1	1	-	-	1	1	-	-	1	1	-	-	1	1	1	-	1	1	1	-	1	-	1
1	-	-	1	1	1	-	1	1	1	1	1	1	-	-	-	-	-	-	1	1	1	1	1	1	-	1	1	1	-	-	1
1	1	1	1	1	-	-	-	-	1	-	-	1	-	1	1	-	-	1	-	1	1	-	1	1	1	1	-	-	-	-	-
1	1	1	1	-	1	-	-	1	-	-	1	-	1	-	1	-	1	-	1	-	1	1	-	1	1	-	1	-	-	-	-
1	1	1	-	1	1	-	1	-	-	1	-	-	1	1	-	1	-	-	1	1	-	1	1	-	1	-	-	1	-	-	-
1	1	-	1	1	1	1	-	-	-	1	1	1	-	-	-	1	1	1	-	-	-	1	1	1	-	-	-	-	1	-	-
1	-	1	1	1	1	1	1	1	1	-	-	-	-	-	-	1	1	1	1	1	1	-	-	-	-	-	-	-	-	1	-
1	1	1	1	1	1	-	-	-	-	-	-	-	-	-	-	-	-	-	-	-	-	-	-	-	-	1	1	1	1	1	1

Figure 1.8 Hadamard matrices *Cl*(1)…*Cl*(5).
Sign signature of Clifford algebras is shown in red.

APPENDIX 1.3

CLIFFORD ALGEBRA *Cl*(4, 2)

Below we present squares of basic elements of Clifford algebra $Cl(4, 2)$, which will be used in the future.

———— **Cl (4, 2), dim = 64, [36+ 28–]**

$e_0^2 = 1.$
$e_1^2 = 1, e_2^2 = 1, e_3^2 = 1, e_4^2 = 1, e_5^2 = -1, e_6^2 = -1.$
$e_{12}^2 = -1, e_{13}^2 = -1, e_{14}^2 = -1, e_{15}^2 = 1, e_{16}^2 = 1, e_{23}^2 = -1, e_{24}^2 = -1,$
$e_{25}^2 = 1, e_{26}^2 = 1, e_{34}^2 = -1, e_{35}^2 = 1, e_{36}^2 = 1, e_{45}^2 = 1, e_{46}^2 = 1,$
$e_{56}^2 = -1.$

— 20 Triplets with signature [12+ 8–]:

$e_{123}^2 = -1, e_{124}^2 = -1, e_{125}^2 = 1, e_{126}^2 = 1, e_{134}^2 = -1, e_{135}^2 = 1,$
$e_{136}^2 = 1, e_{145}^2 = 1, e_{146}^2 = 1, e_{156}^2 = -1, e_{234}^2 = -1, e_{235}^2 = 1,$
$e_{236}^2 = 1, e_{245}^2 = 1, e_{246}^2 = 1, e_{256}^2 = -1, e_{345}^2 = 1, e_{346}^2 = 1,$
$e_{356}^2 = -1, e_{456}^2 = -1.$

—

$e_{1234}^2 = 1, e_{1235}^2 = -1, e_{1236}^2 = -1, e_{1245}^2 = -1, e_{1246}^2 = -1,$
$e_{1256}^2 = 1, e_{1345}^2 = -1, e_{1346}^2 = -1, e_{1356}^2 = 1, e_{1456}^2 = 1,$
$e_{2345}^2 = -1, e_{2346}^2 = -1, e_{2356}^2 = 1, e_{2456}^2 = 1, e_{3456}^2 = 1,$
$e_{12345}^2 = -1, e_{12346}^2 = -1, e_{12356}^2 = 1, e_{12456}^2 = 1, e_{13456}^2 = 1,$
$e_{23456}^2 = 1.$
$e_{123456}^2 = -1.$

————

Relation of revers is used at square calculations:

$$e_{123}^2 = e_{123} \times e_{123} = -e_{123} \times e_{321} = -e_1^2 \times e_2^2 \times e_3^2 = -1;$$
$$e_{2456}^2 = e_{2456} \times e_{6542} = e_2^2 \times e_4^2 \times e_5^2 \times e_6^2 = 1.$$

Product of Clifford algebra elements may be calculated using index permutation in basic elements and finding squares:

$$e_{23} \times e_{134} = -e_{23} \times e_{314} = -e_3^2 \times e_{214} = e_{124};$$
$$e_{245} \times e_{123456} = e_{245} \times e_{234516} = -e_{245} \times e_{245136} =$$
$$= e_{245} \times e_{542} \times e_{136} = e_2^2 \times e_4^2 \times e_5^2 \times e_{136} = -e_{136}.$$

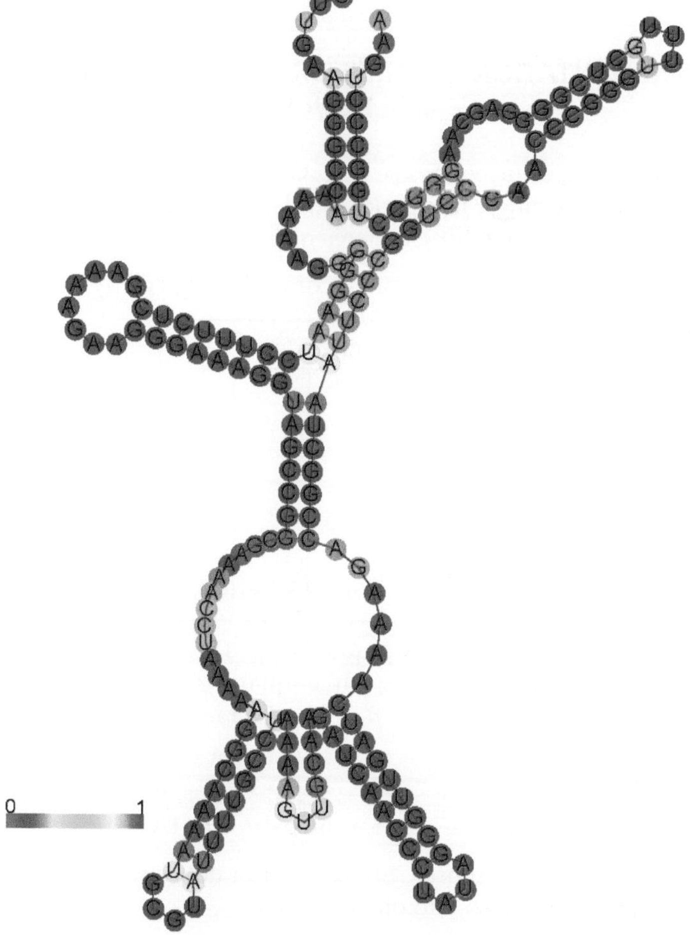

Figure 1.9 RNA structure, which is a realization of Clifford algebra *Cl*(5, 4). The structure was diagrammed using RNA folding program on the site RNAfold WebServer (Institute for Theoretical Chemistry/University of Vienna). MFE structure drawing encoding base-pair probabilities.

APPENDIX 1.4

PRESENTATION OF CLIFFORD ALGEBRA BY RNA MOLECULES

RNA contains single nucleotides **A, U, C, G** and paired nucleotides **AU, UA, CG, GC**, that is – 4 single nucleotides and 4 paired ones. Clifford algebra *Cl*(3) contains eight generating squares of 4-interval.

Signature of generating squares of algebra *Cl*(3) is given by:

	e_0^2	e_1^2	e_2^2	e_3^2
n_1	+	–	–	–
n_2	+	+	–	–
n_3	+	–	+	–
n_4	+	–	–	+
n_5	+	+	+	–
n_6	+	+	–	+
n_7	+	–	+	+
n_8	+	+	+	+

Squares of intervals n_7, n_6, n_5 equal to [3+ 1–] and square n_1 = [1+ 3–] may be kept for single nucleotides, while squares of intervals n_4, n_3, n_2 equal to [2+ 2–] and square of interval n_8 = [4+ 0–] may be reserved for paired nucleotides.

Squares of intervals n_9 = – – + +, n_{10} = – + – +, n_{11} = – + + – equal to [2+ 2–] with $e_0^2 = -1$ may be kept for the pairs **GU, UG**.

The idea is to decompose Clifford algebra not only into two types of Lorentz intervals, but also into other intervals.

Now we will present Clifford algebra as a mix of squares of eight intervals $n_1 \dots n_8$.

The equations for determining numbers $n_1 \dots n_8$ are given by:

$$\{3(\mathbf{P+}) - (\mathbf{Q-})\}/4 = 2n_{765} + 3n_8 + n_{432};$$

$$\{3(\mathbf{Q-}) - (\mathbf{P+})\}/4 = 2n_1 + n_{432} - n_8,$$

where $n_{765} = n_7 + n_6 + n_5$, $n_{432} = n_4 + n_3 + n_2$. Squares of intervals n_9, n_{10}, n_{11} may be included in the set of numbers n_{432}.

These equations are always soluble.

Let us select the coding of nucleotides and their pairs in RNA as follows: **A** = n_1, **G** = n_7, **U** = n_6, **C** = n_5, **AU** = n_8, **UA** = n_4, **GC** = n_3, **CG** = n_2, **GU** = n_9, **UG** = n_{10}.

Further, we will give an example of how to build RNA, which realises Clifford algebra $Cl(5, 4) = [272+\ 240-]$.

One of the solutions of the following equations

$$144 = 2n_{765} + 3n_8 + n_{432};$$
$$112 = 2n_1 + n_{432} - n_8,$$

determines numbers:

$n_1 = 40$, $n_{765} = 30$, that is, 70 single nucleotides, and
$n_8 = 13$, $n_{432} = 45$, that is, 58 paired nucleotides totally.

If numbers $n_1 \ldots n_8$ are taken in excess, excess nucleotides will be neutral towards elements of Clifford algebra.

RNA sequence with calculated numerical parameters may be chosen as follows:

CUUGAAGGGCCAAAAAAGGGGAAUCCUUUCUCGAAAAGAAGGGAAAG
GUAGCCGGCGAAAACCUAAAAAAGGCAAAAUGCGUAUUUUGCCUAAAAG
UUGCAAGAUCAACCCUAUAGGGUUGAUCAAAAGACCGGCUAAUUCCCG
GUCCCAACCCGGGUUUUGCUCGGGGAGCAAGGGCCUGGCCCUGAA

Figure 1.9 shows RNA structure, which is a realization of Clifford algebra $Cl(5, 4)$ with three neutral nucleotides in n_{765}. Such RNA structure has great resemblance to structures of natural RNA.

Selected Bibliography

DNA structure

Saenger, W. Principles of Nucleic Acid Structure. Berlin: Springer-Verlag, 1984.

Singer, M., Berg, P. Genes and genomes. California: University Science Books, 1991.

Bloomfield, V.A., Crothers, D.M., Tinoco, J., Jr. Nucleic acids. Structures, properties, and functions. California: University Science Books, 2000.

Watson, J.D., Baker, T.A., Bell, S.P., Gann, A., Levine, M., Losick, R. Molecular biology of the gene. New York: Cold Spring Harbor Laboratory, 2004.

Clark, D.P. Molecular biology. Understanding the genetic revolution. Elsevier Academic Press, 2005.

Weaver, R.F. Molecular biology. New York: McGraw-Hill, 2012.

Clifford algebra

Shirokov, D.S. Clifford algebras and spinors. Moscow: Steklov Institute of Mathematics, RAS, 2011 (In Russian).

Lounesto, P. Clifford algebras and spinors. Cambridge: Cambridge University Press, 2001.

Gracia-Bondia, J.M., Varilly, J.C., Figuerra, H. Elements of noncommutative geometry. Berlin: Birkhauser, 2000.

Aschieri, P., Dimitrijevic, M., Kulish, P., Lizzi, F., Wess, J. Noncommutative spacetimes. Symmetries in noncommutative geometry and field theory. Berlin Heidelberg: Springer, 2009.

Lorentz group

Gel'fand, I.M., Minlos, R.A. and Shapiro, Z. Ya. Representations of the rotation and Lorentz groups and their applications. New York: Pergamon Press, 1963.

Elliott, J.P., Dawber, P.G. Symmetry in Physics. Volume 2: Further Applications. London: The Macmillian Press, 1979.

Quantum field theory

Ta-Pei Cheng and Ling-Fong Li. Gauge theory of elementary particle physics. Oxford: Clarendon Press, 1984.

Raider, L.H. Quantum Field Theory. Cambridge: Cambridge University Press, 1985.

Quang Ho-Kim, Xuan-Yem Pham. Elementary Particles and their Interections: Concepts and Phenomena. Berlin: Springer-Verlag, 1998.

2

tRNA Molecule in Affine Space

Transport RNA (tRNA) are intermediate links, molecular adapters in the chain of protein synthesis on ribosome.

These small molecules, composed of 74–95 nucleotides, covalently unite anticodons of the genetic code and amino acids. tRNA structure in the form of a clover-leaf (Figure 2.1) is well-known; its tertiary structure resembles the Latin character L.

All tRNA molecules contain many modified nucleotides (frequently—inosine **I**, ribothimidine **T**, pseudouridine **Ψ**, dihydrouridine **D**), which are basically intended for creating standard three-dimensional surface (especially in the anticodon region) necessary for ribosome to recognize them.

tRNA molecules contain four regions, each possessing invariant properties irrespective of the amino acid linked. 3′-End of acceptor stem always contains **CCA** triplet; anticodon is located at the opposite end in position 34–36 of tRNA nucleotides. tRNA tertiary structure is stabilized by hydrogen bonds in the stems.

Standard tRNA structure includes 76 nucleotides. However, D-loop often has an expanded structure; there is also a variable loop with variable number of nucleotides. tRNA molecule contains approximately 20 paired nucleotides, which do not always form canonical pairs.

Often, acceptors of the same amino acid may be several different, isoaccepting tRNA with different anticodons; however, there may be tRNAs with identical anticodons but different structures. The functions of such tRNA are not clear. Nevertheless, surface shape and volume of tRNA depend little on the primary structure.

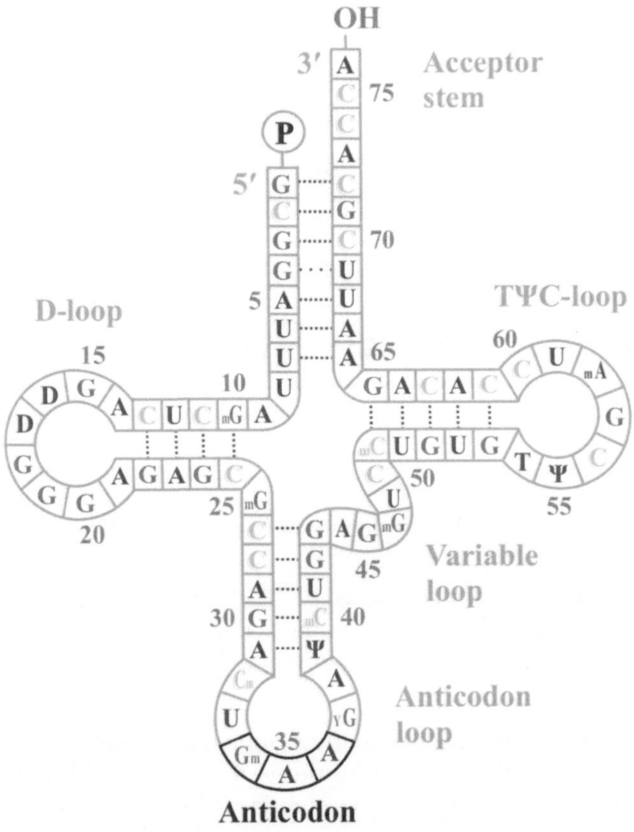

Figure 2.1 Structure of **Phe**-tRNA.

To perform adapter functions in the process of mRNA translation, tRNA molecule should be attached to the amino acid, which corresponds to mRNA codon, since nonloaded tRNA is not perceived by ribosome. Joining of amino acids to tRNA is realized by highly specific enzymes, aminoacyl-tRNA synthetases. Aminoacyl-tRNA synthetases, which are subdivided into two classes, attach amino acid to 2′-OH or 3′-OH end of ribose on the last nucleotide (A76) of tRNA. Further, due to the reaction of transetherification, all tRNAs enter into ribosome in position 3′-OH of attached amino acid. The number of aminoacyl-tRNA synthetases is equal to the number of amino acids (20); they entirely ensure the genetic code.

Ribosome does not react to the amino acid attached to tRNA, but reacts to the absence of an amino acid. We do not know what the reasons for this are.

Much is not yet clear about the behaviour of both tRNA and aminoacyl-tRNA synthetases. Why do only 20 amino acids attach to tRNA, whereas more can be attached artificially? Why are aminoacyl-tRNA synthetases divided into two classes depending on the size of amino acids and their attachment to different ends of ribose? It is not clear why for amino acid **Lys** there are aminoacyl-tRNA synthetases in two classes; what is the peculiarity of amino acid **Phe**, which attaches not to 3′-OH but to 2′-OH end of ribose?

We will try to answer some of these questions.

CLIFFORD ALGEBRAS OF tRNA AND AMINOACYL-tRNA SYNTHETASES

Interaction between molecules may take place on the basis of Clifford algebra. Suppose that a certain molecule M (aminoacyl-tRNA synthetase) interacts with N particles (tRNA nucleotides) of arbitrary nature. We will join these particles into a chain. Displacement of each particle in the chain changes its length by $\delta l_i = \mu_{ik} x^k$, where x^k is four-dimensional coordinates of the knot k in the chain. Index k is double, for example x^{k1}, x^{k2}, x^{k3}, x^{k4}. The second index may be applied to tRNA nucleotides. Quadratic form $\Omega_1 = Z^{ip} \delta l_i \delta l_p = Z^{ip} \mu_{ik} \mu_{pj} x^k x^j$ is reduced by the first index in x^k to the canonical form with p positive and q negative squares, i.e., this form generates Clifford algebra $Cl(p, q)$. The second quadratic form $\Omega_2 = Z^{ip} \mu_i \mu_p$ refers to a surface in the conjugate space (functional) of tRNA on the set of aminoacyl-tRNA synthetases and corresponds to some Clifford algebra.

Pairing of tRNA nucleotides may be considered relatively approximate twofold rotation axis, which separates T- and acceptor branch from D- and anticodon branch. It is impossible to determine precisely the pairs of tRNA nucleotides, symmetrical towards pseudoaxis of second order. Plausible pair numbers is equal to 10. Then, tRNA structure induces two identical Clifford algebras $Cl(10, 10)$: on the part of acceptor stem and on the part of anticodon. Clifford algebra $Cl(10, 10)$ determines the presence of two classes of aminoacyl-tRNA synthetase monomers.

Monomers of aminoacyl-tRNA synthetases of the class I attach all amino acids **R, C, I, L, V, M, Q, E, W, Y** to the 2′-OH end of ribose. At that, Clifford algebra $Cl(0, 10)$ is induced on the assumption that 3′-OH end of ribose determines positive and 2′-OH end—negative part of Clifford algebra signature.

Aminoacyl-tRNA synthetase monomers of the class II attach amino acids **G, H, P, T, S, N, D, K, A** to the 3′-OH end of ribose and amino acid **F** to the 2′-OH end of ribose. At that, Clifford algebra $Cl(9, 1)$ is induced.

We will define amino acids, attached to aminoacyl-tRNA synthetases of the classes I and II, as sets Ψ_I and Ψ_{II}, accordingly. Let us consider Clifford algebras that are induced by sets Ψ_I и Ψ_{II}, taking into account the number of hydrogen bonds of the central codon-anticodon pair of mRNA-tRNA pairing according to the genetic code (Table 2.1).

Table 2.1 The number of hydrogen bonds of the central pair mRNA codon—tRNA anticodon and structure of aminoacyl-tRNA synthetase subunits.

Ψ_I set	R	C	I	L	V	E	Q	W	M	Y
The number of hydrogen bonds	3	3	2	2	2	2	2	3	2	2
Subunits	α	α	α	α	α	α	α	α_2	α_2	α_2
Ψ_{II} set	K	H	P	T	S	N	D	A	G	F
The number of hydrogen bonds	2	2	3	3	3	2	2	3	3	2
Subunits	α_2	α_2	α_2	α_2	α_2	α_2	α_2	α_4	$(\alpha\beta)_2$	$(\alpha\beta)_2$

α – monomer, α_2 – dimer, α_4 – tetramer, $(\alpha\beta)_2$ – heterotetramer.

Suppose that three hydrogen bonds of the central pair mRNA codon—tRNA anticodon determine positive part of induced Clifford algebra signature, while two hydrogen bonds determine negative part. Then, Ψ_I set will induce Clifford algebra $Cl(3, 7)$, while Ψ_{II} will induce Clifford algebra $Cl(5, 5)$. And we have chains of isomorphism of the algebras $Cl(0, 10) \cong Cl(3, 7) \cong Cl(7, 3) \cong Cl(4, 6)$ and $Cl(5, 5) \cong Cl(10, 0) \cong Cl(9, 1) \cong Cl(1, 9) \cong Cl(6, 4)$. We revealed that Clifford algebras, which are determined by aminoacyl-tRNA synthetases for amino acid sets Ψ_I and Ψ_{II}, are isomorphic to Clifford algebras, which are induced by mRNA-tRNA pairing.

The sum of sets Ψ_I and Ψ_{II} is determined by Clifford algebra $Cl(20, 0) \cong Cl(0, 20) \cong Cl(9, 11)$. It is proved experimentally that all amino acids connect to each other through binding to 3′-OH end of tRNA ribose, which conforms to Clifford algebra $Cl(20, 0)$. Induced by acceptor stem Clifford algebra $Cl(10, 10) \cong Cl(11, 9)$ does not unite amino acids into one set, instead it divides them by 10 amino acids in two sets Ψ_I and Ψ_{II}, that are "recognized" by different classes of aminoacyl-tRNA synthetases. Such distribution is based on the fact that affine space has two isomorphic symmetric curvature tensors including each of the 10 components (see further).

Aminoacyl-tRNA synthetase monomers of the class I (Ξ_I) have Rossmann fold in their active centre (5, 6 parallel β-sheets), while aminoacyl-tRNA synthetase monomers of the class II (Ξ_{II}) contain seven antiparallel β-sheets in their active centre. In spite of this, aminoacyl-tRNA synthetase sets of classes I and II are completely equivalent. This is because aminoacyl-tRNA synthetase monomers of the class I may also form dimers, while aminoacyl-tRNA synthetase monomers of the class II form dimers or tetramers.

Dimerization is the arrangement on secondary diagonal of all squares of Clifford algebra basis (including e_0^2) with opposite signs by ordering in opposite way, i.e., Ξ and Ξ^{π}, which is equivalent to multiplication of Clifford algebra bases by imaginary unit $e_k \to ie_k$. Tetramer is a dimer with basis ie_k. Dimers of Clifford algebra $Cl(p, p+1)$ are isomorphic to algebra $Cl(p, p+1)$ itself.

The sets of aminoacyl-tRNA synthetases of the class I and II may be presented as

$$\{\Xi_{I'}, \Xi_I \oplus \Xi_I^{\pi} \sim \Xi_{II}\} \text{ and } \{\Xi_{II} \oplus \Xi_{II}^{\pi} \sim \Xi_{I'}, i\Xi_{II} \oplus (i\Xi_{II})^{\pi} \sim \Xi_{II}\}$$

accordingly. It is clear that these sets are equivalent. That is why lysyl-tRNA synthetases exist in either class; it is only not clear why just for Lys.

Equivalence of aminoacyl-tRNA synthetase sets of classes I and II may be presented as identity of protonic charges for combination of amino acids: $Q_p(\mathbf{W} + \mathbf{M} - \mathbf{Y}) = Q_p(\mathbf{A} + \mathbf{F} - \mathbf{G}) = 57$. Or in detailed record: **5C 12H –1O 1N 1S** ≅ **8C 9H.**

Let us transfer the atom of *anti*-oxygen to the right and reduce the number of hydrogen atoms by two units. We will have the identity:

$$Cl(5\mathbf{C} + 11\mathbf{H} + 1\mathbf{N}, 1\mathbf{S}) = Cl(8\mathbf{C}, 1\mathbf{O} + 8\mathbf{H}) = Cl(48, 16).$$

CURVATURE TENSOR OF FOUR-DIMENSIONAL RIEMANN SPACE

Let us add amino acid **F** to the set $\Psi_{I'}$; we will have the set of amino acids $\Psi_I(+\mathbf{F})$. The set of amino acids Ψ_{II} reduces by amino acid **F:**

$\Psi_{II} \to \Psi_{II}(-\mathbf{F})$. Sets $\Psi_I(+\mathbf{F})$ and $\Psi_{II}(-\mathbf{F})$ are **spinor components of curvature tensor in four-dimensional Riemann space** by the mode of connecting to OH-end of ribose; they determine Clifford algebra $Cl(9, 11)$. This statement is based on the fact that Louis Witten, as far back as 1959, determined two spinors of four-dimensional curvature tensor that contained 11 and 9 components.

In four-dimensional affine space, curvature tensor has 96 components and disintegrates in 5 constituents:

1) 64 components (only 20 ones in Riemann space);
2) two isomorphic symmetrical tensors each containing 10 components—totally 20 components;

3) two isomorphic antisymmetrical tensors each containing 6 components—totally 12 components.

But affine space has no metric, and mRNA translation requires intervals [3+ 1–] of ribosome movement. How can we pass from affine space to the Riemann one?

If amino acids are left in affine space, 20 amino acids will correspond to torsion tensor, which also has 20 components. Possibly, at early evolution of translation apparatus, amino acids were components of torsion tensor in four-dimensional affine space, but, at the present time, Clifford algebra $Cl(6)$ determines distribution of amino acids in curvature tensor of four-dimensional affine space, while amino acids themselves act as spinor components of curvature tensor in Riemann space.

To generate algebra $Cl(6)$, there must be six initial amino acids. The function of Clifford algebra generator may be performed by one of curvature tensors of affine space, which contains six components (generator of left, L-algebra).

Each of the four sections of curvature tensor in four-dimensional Riemann space contain 36 nonzero components, among which there are only 14 independent ones. The rest of the six components should be chosen from other sections: from the second one—5, and from the third one—1. We will define five components from the second section and one component from the third section of Riemann curvature tensor as generating spinors of four-dimensional curvature tensor in affine space (they will be transferred in affine space). This directly determines Clifford algebra with 64 components of curvature tensor in affine space:

$$\boldsymbol{Cl(6, 0)} \cong Cl(1, 5) \cong Cl(5, 1) \cong Cl(2, 4) = [28+\ 36-].$$

All 36 nonzero components of Riemann curvature tensor have negative signature; the rest, arrived from other sections (including generating ones), have positive signature. Six components of Riemann curvature tensor may also be chosen in a different way, but it will not be compact.

CURVATURE TENSOR AND L-FUNCTION OF tRNA

Each component of curvature tensor is mapped on one of 96 tRNA nucleotides. Therefore, tRNA represents a genetic curvature tensor in four-dimensional affine space.

Ten components of symmetrical curvature tensor and six components of antisymmetrical curvature tensor (α-helices) belong to aminoacyl-tRNA synthetases.

To describe the structures of unloaded tRNA, we will consider L-function of sequence of nucleotides $\boldsymbol{n}$, which depends on three indices i, j, k.

Formula for L-function of tRNA is rather cumbersome. Its very beginning components are:

$$\mathrm{L}_{ij}{}^{k} = \mathrm{R}_{ijp}{}^{k}\boldsymbol{n}^{p} + \mathrm{R}_{ij}{}^{kmf}\boldsymbol{n}_{mf} + \mathrm{R}_{ijp}{}^{km}\boldsymbol{n}^{p}\boldsymbol{n}_{m} + \mathrm{R}_{ijp}{}^{kmn}\boldsymbol{n}^{p}\boldsymbol{n}_{m}\boldsymbol{n}_{n} + \ldots$$

Formula for L-function contains separate nucleotides with their indices and paired nucleotides. Each tRNA nucleotide may have one ($\boldsymbol{n}^{p}$), two ($\boldsymbol{n}_{mk}$) or three ($\boldsymbol{n}_{mk}{}^{p}$) indices.

L-function represents the map in bundle of curvature tensor with the use of three indices in four-dimensional affine space.

L-function is the transformation of tRNA for functioning in Riemann space with the use of nucleotide modifications.

Indices i, j, k may change within the ranges:

$i = 1\ldots64$ are numbers of ordinary tRNA nucleotides (**A**, **U**, **C**, **G**) that represent 64 components of curvature tensor;

$j = 1\ldots20$—numbers of nucleotides of tRNA structure expansion, which map two isomorphic symmetrical curvature tensors;

$k = 1\ldots12$—numbers of modified tRNA nucleotides that represent two isomorphic antisymmetrical curvature tensors.

L-function always depends on three nucleotides. The first index points to usual nucleotide, the second index points to the nucleotide number responsible for the structure expansion, and the third index points to the nucleotide modification number.

For definiteness, we will suppose that the first index of tRNA nucleotide designates its arrangement along the chain (1…64): lower index—nucleotide is located before the centre and in the centre of anticodon; upper index—nucleotide is located after the centre of anticodon. The second index of tRNA nucleotide designates expansion of the standard nucleotide sequence (1…20); location of the second index coincides with location of the first one. The third index is always oriented oppositely to the first one and designates various chemical modification of the nucleotide (1…12).

Suppose that nucleotides of variable chain belong to nucleotides of expansion or to modified nucleotides.

There are difficulties in determining status of such nucleotides as **T**, **D**, **I**, **Ψ** that may be interpreted as standard ones.

If the number of nucleotides of structure expansion is more than 20, part of them must be added to the standard nucleotides. If list of modified nucleotides numbers more than 12, part of them must be counted among the standard nucleotides or the nucleotides of expansion.

Each nucleotide contains first index. If second index is absent, index of modification is always placed opposite to the first one and is located in the place of the second index.

Modified nucleotide, which expands tRNA structure, always has three indices. The record $\boldsymbol{n}_{20,1}{}^{5}$ stands for the following: first expansion after twentieth standard nucleotide with number 5 in the list of modified nucleotides for concrete tRNA.

Note that every modified nucleotide, which has three indices, is a strongly compressed map (retractor) of all tRNA structure, since $\mathrm{L}_{ij}{}^{k}$ is similar to $\boldsymbol{n}_{sp}{}^{m}$.

If a nucleotide contains parts of indices of anticodon or acceptor stem, this nucleotide is associated with tRNA recognition by aminoacyl-tRNA synthetase.

Let us write out characteristic components of tRNA L-function. The rest of the components may be obtained by combining nucleotide indices and transposing nucleotide index position.

Summation indices in R-functions possess the same values as indices of nucleotides.

1. **Terms with one nucleotide.** These nucleotides, disposed in tRNA loops, are summarized by R-functions that possess up to 6 indices:

$$\mathrm{R}_{ijm}{}^{k}\boldsymbol{n}^{m},\ \mathrm{R}_{ij}{}^{kmp}\boldsymbol{n}_{mp},\ \mathrm{R}_{ijs}{}^{kmp}\boldsymbol{n}_{mp}{}^{s}.$$

2. **Terms with two nucleotides.** Nucleotide pairs (not obligatory complementary-paired) are summarized by R-functions that possess up to 9 indices:

$$\mathrm{R}_{ijm}{}^{kp}\boldsymbol{n}_{p}\boldsymbol{n}^{m},\ \mathrm{R}_{ij}{}^{kmps}\boldsymbol{n}_{mp}\boldsymbol{n}_{s},\ \mathrm{R}_{ijsm}{}^{kp}\boldsymbol{n}_{p}{}^{s}\boldsymbol{n}^{m},$$
$$\mathrm{R}_{ijrm}{}^{ksp}\boldsymbol{n}^{r}\boldsymbol{n}_{sp}{}^{m},\ \mathrm{R}_{ijrsm}{}^{kp}\boldsymbol{n}^{rs}\boldsymbol{n}_{p}{}^{m},\ \mathrm{R}_{ijsm}{}^{kftpr}\boldsymbol{n}_{ft}{}^{s}\boldsymbol{n}_{pr}{}^{m}.$$

3. **Terms with three nucleotides.** Three nucleotides (not only anticodon or located next nucleotides) are summarized by R-functions possessing up to 11 indices:

$$\mathrm{R}_{ijmn}{}^{kp}\boldsymbol{n}_{p}\boldsymbol{n}^{m}\boldsymbol{n}^{n},\ \mathrm{R}_{ijmn}{}^{kps}\boldsymbol{n}_{ps}\boldsymbol{n}^{m}\boldsymbol{n}^{n},\ \mathrm{R}_{ijmn}{}^{kpst}\boldsymbol{n}_{ps}\boldsymbol{n}^{m}{}_{t}\boldsymbol{n}^{n},$$
$$\mathrm{R}_{ijsmn}{}^{kftpr}\boldsymbol{n}_{ft}{}^{s}\boldsymbol{n}_{pr}{}^{m}\boldsymbol{n}^{n},\ \mathrm{R}_{ismnx}{}^{kftpr}\boldsymbol{n}_{ft}{}^{s}\boldsymbol{n}_{pr}{}^{m}\boldsymbol{n}^{nx}{}_{j},$$
$$\mathrm{R}_{ijsmnx}{}^{kftpr}\boldsymbol{n}_{ft}{}^{s}\boldsymbol{n}_{pr}{}^{m}\boldsymbol{n}^{nx}.$$

Triplet $\boldsymbol{n}_{ft}{}^{s}\boldsymbol{n}_{pr}{}^{m}\boldsymbol{n}^{nx}{}_{j}$ contains 9 indices, one of which coincides with the index in L-function.

4. **Terms with four nucleotides.** Groups of four nucleotides, randomly disposed along tRNA chain, are summarized by R-functions possessing up to 11 indices:

$$\mathrm{R}_{ijmns}{}^{kp}\boldsymbol{n}_{p}\boldsymbol{n}^{m}\boldsymbol{n}^{n}\boldsymbol{n}^{s},\ \mathrm{R}_{ijmnt}{}^{kps}\boldsymbol{n}_{ps}\boldsymbol{n}^{m}\boldsymbol{n}^{n}\boldsymbol{n}^{t},\ \mathrm{R}_{ijmng}{}^{kpst}\boldsymbol{n}_{ps}\boldsymbol{n}^{m}{}_{t}\boldsymbol{n}^{n}\boldsymbol{n}^{g},$$
$$\mathrm{R}_{ijsmng}{}^{kftpr}\boldsymbol{n}_{ft}{}^{s}\boldsymbol{n}_{pr}{}^{m}\boldsymbol{n}^{n}\boldsymbol{n}^{g},\ \mathrm{R}_{ismnxg}{}^{kftpr}\boldsymbol{n}_{ft}{}^{s}\boldsymbol{n}_{pr}{}^{m}\boldsymbol{n}^{nx}{}_{j}\boldsymbol{n}^{g},$$
$$\mathrm{R}_{ismnxg}{}^{kftpr}\boldsymbol{n}_{ft}{}^{s}\boldsymbol{n}_{pr}{}^{m}\boldsymbol{n}^{n}{}_{j}\boldsymbol{n}^{xg},\ \mathrm{R}_{ismnxg}{}^{ftpr}\boldsymbol{n}_{ft}{}^{s}\boldsymbol{n}_{pr}{}^{m}\boldsymbol{n}^{nx}{}_{j}\boldsymbol{n}^{gk}.$$

Quadruplet $\boldsymbol{n}_{ft}{}^{s}\boldsymbol{n}_{pr}{}^{m}\boldsymbol{n}^{nx}{}_{j}\boldsymbol{n}^{gk}$ contains 11 indices, two of which coincide with the indices in L-function.

Division of curvature tensor spinors of Riemann space into 11 and 9 components is twice fixed in tRNA:

by the set of R-functions, which expresses activity in Riemann space;

by volume structure (L-function—form of the fourth order) that depends on modified nucleotides.

Without modified nucleotides, only R-functions with six indices would be in the composition of tRNA L-function, i.e., tRNA could function only in affine space.

Indices of tRNA L-function take large values, which indicates that this function has nonlocal character. If indices *i*, *j*, *k* were changing in the limits 1…4, tRNA molecule would be a material content of torsion tensor or connection in Riemann space. But that is not so. Indeed, tRNA molecule is a binding molecule; but, remaining in affine space, it unites codons of genetic code, amino acids and aminoacyl-tRNA synthetases, whereas usual connection may unite only two directions.

tRNA, loaded with amino acid A_m (m = 1…20), may be considered as conditional curvature tensor $\rho_{ij}{}^{k}{}_{m} = \mathrm{L}_{ij}{}^{k}\mathrm{A}_{m}$ that contains four indices. Notation $\rho_{ij}{}^{k}{}_{m}$ implies independence of amino acid from tRNA.

HIDDEN SYMMETRY OF THE GENETIC CODE

Informational relationship between mRNA nucleotide sequences and amino acid sequences of proteins is realized by means of the genetic code. Except for minor variations, the genetic code is universal, that is, it is identical for all living organisms (Figure 2.2).

1 \ 2	U	C	A	G	3
U	Phe	Ser	Tyr	Cys	U
	Phe	Ser	Tyr	Cys	C
	Leu	Ser	Stop	Stop	A
	Leu	Ser	Stop	Trp	G
C	Leu	Pro	His	Arg	U
	Leu	Pro	His	Arg	C
	Leu	Pro	Gln	Arg	A
	Leu	Pro	Gln	Arg	G
A	Ile	Thr	Asn	Ser	U
	Ile	Thr	Asn	Ser	C
	Ile	Thr	Lys	Arg	A
	Met	Thr	Lys	Arg	G
G	Val	Ala	Asp	Gly	U
	Val	Ala	Asp	Gly	C
	Val	Ala	Glu	Gly	A
	Val	Ala	Glu	Gly	G

Figure 2.2 Modern table of the genetic code.

In the genetic code, each triplet of mRNA nucleotides (codon) is responsible only for one amino acid. But not vice versa, since the genetic code is degenerated. Besides a simple degeneration **U/C** in third position of codon, there are no other obvious properties of symmetry in the genetic code.

Search for regularities in the genetic code is rather intensive. There was suggested a model of coevolution of amino acids and their conformity with tRNA codons.

Mathematical analysis based on group theory can only partially explain the distribution of amino acids on mRNA codons.

The central idea of such approaches is spontaneous violation of initial symmetry with simultaneous expansion of amino acid number. The cause of such spontaneous symmetry violation is unknown—the source of chaos is not clear.

Much attention has been given to the topological basis of the genetic code. Sense of such constructs is that topology of mRNA nucleotide triplets should determine algebra of amino acid degeneracy. It is obvious that 64 triplets of mRNA may be presented as six-dimensional hypercube.

But it fails to obtain numbers of amino acid degeneracy in this way. The situation is somewhat better with using colour graphs. Taking into account **U/C** degeneracy in third codon position, we may suppose that there are only 48 codon combinations. The genetic code arose as a phase transition in informational channel of nucleotide transformation into amino acids. Matrix of links between codons is a three-dimensional graph that may be inserted into the surface of the genus $\gamma = 25$ and coloured with 20 colours, which number is equal to the number of amino acids.

The reading operator has been proposed, which, acting on the codons, describes the genetic code. Reading operator is similar to projection operators, as the same amino acid may correspond to different codons. Limitation of this approach is that it does not take into account the possibility of linking false amino acid to tRNA.

In order to reveal the code of amino acid interactions, G.I. Chipens (1991) suggested the division of the set of amino acids into three groups depending on the central pair of codon-anticodon triplets. Code of effective interactions between amino acids in the folding of proteins has not so far been found, however, subdividing of amino acids into independent groups seems to be rather important. We use in explicit form the table of the genetic code, constructed on the basis of codon-anticodon coupling of amino acids.

Quadratic form of the genetic code

Figure 2.3 shows the table of the genetic code in the form of amino acid pairs that correspond to mRNA-tRNA codon-anticodon pairing of nucleotides.

Am5*={F,K,E,*,L,Q}	Am8*={S,R,G,*,P,W,T,C,A}
1F(UUU) = K(AAA)1	1S(UCU) = R(AGA)1
1F(UUC) = E(GAA)1	1S(UCC) = G(GGA)1
1L(UUA) = *(UAA)1	1S(UCA) = *(UGA)1
1L(UUG) = Q(CAA)1	1S(UCG) = R(CGA)1
1L(CUU) = K(AAG)1	1P(CCU) = R(AGG)1
1L(CUC) = E(GAG)1	1P(CCC) = G(GGG)1
1L(CUA) = *(UAG)1	1P(CCA) = W(UGG)1
1L(CUG) = Q(CAG)1	1P(CCG) = R(CGG)1
Am7={I,N,D,Y,M,H,V}	
1I(AUU) = N(AAU)1	1T(ACU) = S(AGU)1
1I(AUC) = D(GAU)1	1T(ACC) = G(GGU)1
1I(AUA) = Y(UAU)1	1T(ACA) = C(UGU)1
1M(AUG) = H(CAU)1	1T(ACG) = R(CGU)1
1V(GUU) = N(AAC)1	1A(GCU) = S(AGC)1
1V(GUC) = D(GAC)1	1A(GCC) = G(GGC)1
1V(GUA) = Y(UAC)1	1A(GCA) = C(UGC)1
1V(GUG) = H(CAC)1	1A(GCG) = R(CGC)1

Figure 2.3 Table of the genetic code with the number of tRNA isoforms equal to 1. Three sets of amino acids Am5*, Am7 and Am8* form independent families.

Codons corresponding to terminators generate formal amino acid pairs. We will present families of amino acid pairs by quadratic forms, whose coefficients are equal to the sum of numbers of tRNA anticodon isoforms.

Every quadratic form corresponds to some type of symmetry. For amino acid families, this is just hidden symmetry of the genetic code.

Quadratic forms for three amino acid families are as follows:

Am5* [2+ 2–]: +2FK +2FE +4L* +4LQ +2LK +2LE.
Am7 [2+ 2–]: +2IN +2ID +2IY +2MH +2VN +2VD +2VY +2VH.
Am8* [3+ 3–]: +4SR +2SG +2S* +4PR +2PG +2PW +2TS +2TG +
+2TC +2TR +2AS +2AG +2AC +2AR.

Signatures of quadratic forms are shown in brackets. Terminators in the families Am5* and Am8* may be supposed to be independent, since the sum of these sets has a quadratic form with signature [5+ 5–].

The number of anticodon isoforms in all organisms, beginning from protozoan, is more than one. Anticodon statistics do not influence the signature of quadratic form of every amino acid family Am5*, Am7 or Am8*.

Figure 2.4 presents the table of genetic code in the form of amino acid pairs that correspond to tRNA isoforms for the organism *Danio rerio (Zebrafish).*

Am5*={F,K,E,*,L,Q}	Am8U={S,R,G,U,P,W,T,C,A}
0565F(UUU) = K(AAA)0010	0264S(UCU) = R(AGA)0359
0233F(UUC) = E(GAA)0220	0040S(UCC) = G(GGA)0002
0007L(UUA) = *(UAA)0061	0014S(UCA) = U(UGA)0213
0110L(UUG) = Q(CAA)0182	0104S(UCG) = R(CGA)0092
1018L(CUU) = K(AAG)0306	0101P(CCU) = R(AGG)0346
0238L(CUC) = E(GAG)0004	0279P(CCC) = G(GGG)0002
0009L(CUA) = *(UAG)0270	0038P(CCA) = W(UGG)0226
0297L(CUG) = Q(CAG)0314	0059P(CCG) = R(CGG)0209
Am7={I,N,D,Y,M,H,V}	
0025I(AUU) = N(AAU)0285	0014T(ACU) = S(AGU)0384
0001I(AUC) = D(GAU)0009	0008T(ACC) = G(GGU)0010
0005I(AUA) = Y(UAU)0073	0009T(ACA) = C(UGU)0303
0011M(AUG) = H(CAU)0612	0191T(ACG) = R(CGU)0070
1138V(GUU) = N(AAC)0247	0436A(GCU) = S(AGC)0090
0166V(GUC) = D(GAC)0005	0826A(GCC) = G(GGC)0004
0225V(GUA) = Y(UAC)0243	0098A(GCA) = C(UGC)0367
0405V(GUG) = H(CAC)0235	0025A(GCG) = R(CGC)0112

Figure 2.4 Table of the genetic code with the number of tRNA isoforms for the organism *Danio rerio (Zebrafish)*. The numeric data are taken from *Genomic tRNA Database*. Amino acid **Sec** is conditionally marked as **U**.

Quadratic forms for three amino acid sets (Figure 2.4) are as follows:

Am5* [2+ 2–]: +575FK +453FE +68L* +292LQ +1324LK +242LE +279L* + +611LQ.

Am7 [3+ 3–]: +310IN +10ID +78IY +623MH +1385VN +171VD +468VY + +640VH.

Am8* [4+ 4–]: +623SR +42SG +227SU +196SR +447PR +281PG +264PW + +268PR +398TS +18TG +312TC +261TR +526AS +830AG +465AC +137AR.

Every quadratic form of the signature [p+ q–] corresponds to Clifford algebra $Cl(p, q)$, which has $2^{(p+q)}$ elements. Quadratic form [1+ 1–] will be named 2-interval (double-point set). **For the genetic code with the number of tRNA isoforms equal to 1, the number of 2-intervals in the sets Am5*, Am7 and Am8* coincides with the number of hydrogen bonds in the central complementary pair of codon-anticodon nucleotides.** *The number of 2-intervals will be called the state of amino acid family.*

It is clear that the availability of tRNA isoforms changes the states of amino acid families Am7 and Am8*. However, it is unknown, if this has an effect on ribosome behaviour at mRNA translation.

Clifford algebras $Cl(1,1), Cl(2,2), Cl(3,3), Cl(4,4)$ are hidden symmetries of the genetic code. There is still no answer to the question as to if these amino acid algebras are preserved in the folding of protein molecule.

With regard to tRNA isoforms but ignoring wobble-pairing, hidden symmetry of genetic code may be represented as:

$$Cl(10, 10) \cong Cl(11, 9) \cong Cl(1, 1) \otimes Cl(2, 2) \otimes Cl(3, 3) \otimes Cl(4, 4).$$

Algebra $Cl(11, 9) \cong Cl(19, 1)$ with selected amino acid (possibly **Met**) is a **dimer** of algebra $Cl(9, 11) \cong Cl(1, 19)$ with selected amino acid (possibly **Lys**).

As was mentioned, Clifford algebra $Cl(10, 10)$ divides amino acids into two sets Ψ_I and Ψ_{II}, each containing 10 amino acids, which may be "recognized" by different classes of aminoacyl-tRNA synthetases.

Graph of genetic code states

Figure 2.5 shows graph of the genetic code with hidden (but reversible) functions of transitions between states of different families of amino acids.

The number of transition function states is determined as a set of all maps of one set into another one. The number of state transition functions must be equal to 64, that is, to the number of genetic code codons. Only in that case one may talk about hidden symmetry of the genetic code.

There are no transitions within the amino acid families with the exception of substitution for the triplet **UGA**: from codon of terminator to the codon of selenocysteine.

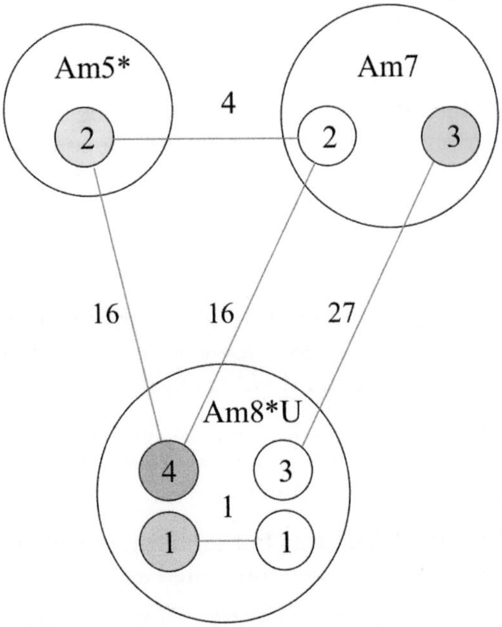

Figure 2.5 The graph of genetic code states. The number of 2-intervals determines the states of amino acid sets. The number of functions of transition states are shown above the lines of connections. Colourless states correspond to amino acid families with the number of tRNA isoforms equal to 1.

Transition 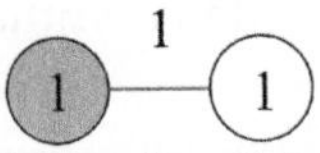

designates the transition of amino acid pairs **SU** ↔ **S***, which implies "elongation of the unit".

Note that there may be many graphs built with irreversible transitions between the states of amino acid families. For instance, the transitions 1 ↔ 1 and 2 ↔ 2 may be joined into one transition 1 ↔ 4. Certainly, the transition 3 → 4 in the family Am8* gives all 64 states of the genetic code. But is it hidden? It is possible, that this transition is connected with wobble-hypothesis: complementary mRNA-tRNA pairs do not form four hydrogen bonds. **An important property of reversible hidden transitions between amino acid families is doubling of genetic code states.**

For all that, the problem of transferring Clifford algebras of amino acid families on amino acid symmetry in the folding of protein molecule, still remains a complex one.

Interpretation of genetic code graph

Authors of research papers on genetic code structure were mainly occupied with the relations between triplets of mRNA nucleotides and amino acids. However, it is clear that the genetic code proceeds further with DNA replication and transcription. Hidden symmetry of the genetic code is linked with these important biological processes.

Let us consider DNA replication: on the graph of genetic code states, there is a triangular image $4 \leftrightarrow (2 \leftrightarrow 2)$. We will present DNA as a complementary vector, consisting of basis vectors e_i in linear (not quadratic) record, which are numbered in the direction **5′3′** of (+)-DNA and **3′5′** of (–)-DNA as follows:

(+)-DNA:	$\mathbf{A}(+e_1)$	$\mathbf{T}(+e_2)$	$\mathbf{C}(+e_3)$	$\mathbf{G}(+e_4)$
(–)-DNA:	$\mathbf{T}(-e_1)$	$\mathbf{A}(-e_2)$	$\mathbf{G}(-e_3)$	$\mathbf{C}(-e_4)$

where (+)-DNA and (–)-DNA are complementary DNA chains.

If we assume that

$$\mathbf{A}(+e_1) = \mathbf{T}(-e_1), \quad \mathbf{T}(+e_2) = \mathbf{A}(-e_2),$$
$$\mathbf{C}(+e_3) = \mathbf{G}(-e_3), \quad \mathbf{G}(+e_4) = \mathbf{C}(-e_4)$$

and also for all nucleotides $\mathrm{N}(+e_1) = +\mathrm{N}e_1$, $\mathrm{N}(-e_1) = -\mathrm{N}e_1$, complementary chains will be identical. If $e_ie_i = 0$ and $e_ie_j = -e_je_i$, complementary chains will be orthogonal. Vectors e_i obey Grassmann algebra. The relationship between Grassmann algebra and Clifford algebra is well known.

If the direction of basis vectors in (+)-DNA chain is chosen randomly, it will give 16 variants of their placement, i.e., transition $4 \rightarrow 2$. At that, by the change of the vector orientation (transition $2 \rightarrow 4$), (–)-DNA chain will conform to (+)-chain. (–)-DNA chain may be considered analogously.

Additionally, 4 functions of transition may be obtained at assumption of possible variants of DNA nucleotide pairing of the following type:

(+)-DNA:	$\mathbf{A}(+e_1)$	$\mathbf{A}(-e_2)$	$\mathbf{A}(+e_3)$	$\mathbf{A}(-e_4)$
(–)-DNA:	$\mathbf{T}(+e_1)$	$\mathbf{T}(-e_2)$	$\mathbf{T}(-e_3)$	$\mathbf{T}(+e_4)$.

It is much more difficult to obtained (27 + 1) transition functions. These functions correspond to DNA nucleotide reading, that is, to DNA transcription. At that, one nucleotide is leading and specifies the way of reading, e.g., **A**. Accordingly, **T**$(-e_1) \rightarrow$ **T**$(-e_1') \rightarrow$ **U**$(+e_1')$, where stroke character marks transition to dual basis. This is connected with Lorentz intervals of nucleotide reading. If in dual basis any triplet of these nucleotides corresponds to every **AGC** triplet, we will obtain 27 transition functions.

The result of (–)-DNA chain transcription is (+)-RNA chain with formal Lorentz metric:

(–)-DNA:	**T**$(-e_1)$	**A**$(-e_2)$	**G**$(-e_3)$	**C**$(-e_4)$
(+)-RNA:	**A**$(+e_1')$	**T**$(-e_2')$	**C**$(+e_3')$	**G**$(+e_4')$.

The following relations are satisfied:

$$\mathbf{A}(+e_1') = \mathbf{T}(-e_1) = \mathbf{A}(+e_1), \quad \mathbf{C}(+e_3') = \mathbf{G}(-e_3) = \mathbf{C}(+e_3),$$

$$\mathbf{G}(+e_4') = \mathbf{C}(-e_4) = \mathbf{G}(+e_4), \quad \mathbf{T}(-e_2') = \mathbf{A}(-e_2) = \mathbf{T}(+e_2).$$

This means that basis direction for nucleotide **T** in (+)-RNA chain should be inverted. It was shown experimentally that **T**$(-e_2')$ = **U**$(+e_2')$ and basis direction for nucleotide **U** coincides with basis direction for other nucleotides.

Transition **T** $\rightarrow$ **U** and $-e_2' \rightarrow +e_2'$ corresponds to mapping **1** $\leftrightarrow$ **1** on the graph of genetic code states.

Let us specify the algebra of DNA transcription more precisely.

One more local basis of Grassmann algebra is introduced with generators a, a^*: $aa^* + a^*a = 0$.

In mRNA, correspondence is established between RNA code and arrangement of basis elements of Grassmann algebra. For virtual RNA with DNA nucleotides we have:

	+A	**+C**	**+G**	**+T**
(+)-RNA:	$+\mathrm{A}_{00}$	$+\mathrm{A}_{10}a^*$	$+\mathrm{A}_{01}a$	$+\mathrm{A}_{11}a^*a$
(–)-RNA:	$-\mathrm{A}_{11}$	$-\mathrm{A}_{01}a$	$-\mathrm{A}_{10}a^*$	$-\mathrm{A}_{00}aa^*$
	–T	**–G**	**–C**	**–A**

Changing of virtual mRNA basis from (+)-chain to (–)-chain requires adjoining in Grassmann algebra, that is $a \rightarrow a^*$, $a^* \rightarrow a$, $aa^* \rightarrow a^*a$. This basis will not be preserved in mRNA.

RNA = read DNA – integral in Grassmann algebra:

(+)-mRNA: K_{00} $\quad \mathrm{K}_{10}a^*$ $\quad \mathrm{K}_{01}a$ $\quad \mathrm{K}_{11}a^*a$

Read RNA: $\mathrm{A}(a^*, a) = e^{a^*a}\,\mathrm{K}(a^*, a)$; where $\mathrm{K}(a^*, a)$ is RNA code.

It is possible that factor of integration e^{a^*a} in Grassmann algebra corresponds to mRNA cap. Relation of RNA with DNA is determined by the formulae of integration

$$K_{00} = A_{00} = \mathbf{A}, K_{10} = A_{10} = \mathbf{C},$$
$$K_{01} = A_{01} = \mathbf{G}, K_{11} = \mathbf{U} = A_{11} - A_{00} = \mathbf{T} - \mathbf{A}$$

and is associated with formal Lorentz metric.

Formulae of integration in Grassmann algebra are presented in a well known book on field theory by authors L.D. Faddeev and A.A. Slavnov (1991).

Basis of mRNA before splicing:

(+)-mRNA:	$+\mathbf{A}$	$+\mathbf{C}a^*$	$+\mathbf{G}a$	$+\mathbf{U}a^*a$
(–)-mRNA:	$-\mathbf{U}aa^*$	$-\mathbf{G}a$	$-\mathbf{C}a^*$	$-\mathbf{A}$.

For the most part, introns at mRNA splicing have the form of lariat that is, in our opinion, the image of sugar-phosphate basis of mRNA. Therefore, **information on the sugar-phosphate basis is cut out from mRNA at splicing**.

Chains of mRNA will be orthogonal, if their product is equal to zero:

$$+\{-\mathbf{AA}\}+\{-\mathbf{AG} - \mathbf{GA}\}a + \{-\mathbf{AC} - \mathbf{CA}\}a^* +$$
$$+\{+\mathbf{AU} - \mathbf{UA} + \mathbf{GC} - \mathbf{CG}\}a^*a = 0.$$

mRNA nucleotides, as operators, satisfy the relations:

$$\mathbf{AA} = 0,\ \mathbf{AG} = -\mathbf{GA},\ \mathbf{AC} = -\mathbf{CA},\ \mathbf{AU} = \mathbf{UA},\ \mathbf{GC} = \mathbf{CG}$$

and form some superalgebra. This basis may be considered as an autocatalytic one in mRNA, but it is not always preserved.

For mRNA after splicing, local 4-basis of Grassmann algebra may be introduced, which depends on spatial variables along the chain:

(+)-mRNA:	$+\mathbf{A}e_1$	$+\mathbf{C}e_2$	$+\mathbf{G}e_3$	$+\mathbf{U}e_4$
(–)-mRNA:	$-\mathbf{U}e_1$	$-\mathbf{G}e_2$	$-\mathbf{C}e_3$	$-\mathbf{A}e_4$

with identities $\mathbf{A} = -\mathbf{U}$, $\mathbf{C} = -\mathbf{G}$, $\mathbf{G} = -\mathbf{C}$, $\mathbf{U} = -\mathbf{A}$. New basis of Grassmann algebra is the final one and is preserved for mRNA.

Therefore, for DNA replication and transcription, we mapped all the sets of transition functions on the graph of genetic code states into the set of genetic information transformations.

Thus, the genetic code may be subdivided into three sets with hidden symmetry (Clifford algebras). Genetic code structure is of the type that it has hidden connection with the processes of DNA replication and transcription.

DNA nucleotide basis is linearly ordered global basis of Grassmann algebra. For informational nucleotides in RNA, local basis of Grassmann algebra is present, and this basis changes at RNA transformation. At splicing, introns are cut out from initial RNA transcripts, that, possibly, may encode sugar-phosphate basis of RNA.

DEGENERACY OF THE GENETIC CODE

The genetic code, expressing dependence of amino acids on mRNA triplets, is degenerated. This means that amino acids, except methionine and tryptophan, are coded by more than one nucleotide triplet. From amino acid sequence, it is impossible to uniquely restore the sequence of mRNA coding triplets.

Let us assume that origin of the genetic code is closely associated with Riemann geometry. Twenty amino acids of the genetic code correspond exactly to the number of independent components of curvature tensor of four-dimensional space-time, which just determines degeneracy of the genetic code.

Matrix RNA has a complex three-dimensional structure. Therefore, mRNA nucleotides are located in curved frame of reference relative to the sugar-phosphate basis. At translation of mRNA triplets into genetic code amino acids, the dynamics of triplets to amino acids transformation should be taken into account. Therefore, we may assume that the process mRNA triplet translation into amino acids takes place in curved four-dimensional space-time. This curved space is Riemann space of coding mRNA nucleotides.

Let us denote by ξ arbitrary coding nucleotide of mRNA or tRNA. Then, variation of mRNA nucleotide ξ at its transfer along ribosome may be determined on the basis of the connecting formula of Riemann geometry:

$$\delta\xi_i = \Gamma^k{}_{it}\,\xi_k\, dx^t, \tag{1}$$

where

$\Gamma^k{}_{it}$ is space-time connection of mRNA triplet $\boldsymbol{kit}$;
x^t is space-time coordinates of mRNA nucleotide ξ_t.

Variation of tRNA nucleotide $\delta\xi^i$ is determined analogously:

$$\delta\xi^i = -\Gamma^i{}_{\alpha\beta}\,\xi^\alpha\, dx^\beta, \tag{2}$$

where

$\Gamma^i{}_{\alpha\beta}$ is space-time connection of anticode tRNA triplet $\boldsymbol{i\alpha\beta}$;
x^β is space-time coordinate of tRNA nucleotide ξ^β.

Let us multiply (1) and (2). This will result in square of invariant space-time interval ds^2 of the change of coding mRNA nucleotide:

$$ds^2 = \delta\xi_i \delta\xi^i = -\Gamma^k_{\ it}\Gamma^i_{\ \alpha\beta}\,\xi_k\,\xi^\alpha dx^t dx^\beta = g_{t\beta}\,dx^t dx^\beta,$$

where $g_{t\beta}$ is the metric tensor of four-dimensional Riemann space.

Let us present the product $\xi_k\,\xi^\alpha$ in two ways:

$$\xi_k\,\xi^\alpha = g_{k\nu}\,\xi^\nu\,\xi^\alpha = g^{\nu\alpha}\,\xi_k\,\xi_\nu.$$

Hypersurface of mRNA

$$a^{en}\xi_e\xi_n + a^\nu\xi_\nu = 0$$

can be defined as a solution of mutually dual equations

$$\xi_\nu = S_\nu^{\ en}\,\xi_e\xi_n \qquad (3)$$

and

$$a^{en} = -S_\nu^{\ en}\,a^\nu.$$

The first equation completely determines the second one, which may be the equation of some factor algebra.

We will present the tRNA nucleotide product $\xi^\nu\xi^\alpha$ through amino acid vector ζ^μ:

$$\xi^\nu\xi^\alpha = Z_\mu^{\ \nu\alpha}\,\zeta^\mu. \qquad (4)$$

Note that relation (4) does **not determine any algebra.**

For arbitrary algebra, there should be fulfilled the relation:

$$\eta^\nu\eta^\alpha = \gamma_\mu^{\ \nu\alpha}\,\eta^\mu,$$

where $\gamma_\mu^{\ \nu\alpha}$ are structural constants of the algebra.

Necessity to introduce relation (4) is connected with the irreversibility of mRNA triplet transformation into amino acid.

Relations (3) and (4) result in dependence of coding mRNA nucleotides on amino acids, not vice-versa:

$$\xi_\nu = S_\nu^{\ en}\,\xi_e\xi_n = g_{e\chi}\,g_{n\alpha}\,S_\nu^{\ en}\,\xi^\chi\,\xi^\alpha = g_{e\chi}\,g_{n\alpha}\,S_\nu^{\ en}\,Z_\mu^{\ \chi\alpha}\,\zeta^\mu.$$

In terms of the relations (3) and (4), we will get two representations of the metric tensor $g_{t\beta}$:

$$\begin{aligned} g_{t\beta} &= -\Gamma^k_{\ it}\Gamma^i_{\ \alpha\beta}\,g^{\nu\alpha}\,\xi_k\,\xi_\nu = \\ &= -\Gamma^k_{\ it}\Gamma^i_{\ \alpha\beta}\,g^{\nu\alpha}\,S_\nu^{\ en}\xi_k\xi_e\xi_n = \Omega_{t\beta}^{\ \ ken}\,\xi_k\xi_e\xi_n. \end{aligned} \qquad (5)$$

$$\begin{aligned} g_{t\beta} &= -\Gamma^k_{\ it}\Gamma^i_{\ \alpha\beta}\,g_{k\nu}\,\xi^\nu\,\xi^\alpha = \\ &= -\Gamma^k_{\ it}\Gamma^i_{\ \alpha\beta}\,g_{k\nu}\,Z_\mu^{\ \nu\alpha}\,\zeta^\mu = \Xi_{\mu t\beta}\,\zeta^\mu. \end{aligned} \qquad (6)$$

Let us determine symmetric differences of right parts of the relations (5) and (6):

$$\Omega_{t\beta}^{\ ken} (-) \Xi_{\mu t\beta} = R^{\ ken}_{\ \ \mu}; \ \xi_k \xi_e \xi_n (-) \zeta^{\mu} = \Lambda_{ken}^{\ \ \mu}.$$

The correspondence of amino acids ζ^{μ} to triplets $\xi_k\xi_e\xi_n$ is single-valued; their intersection region is a *terminator* set.

The function $R^{ken}_{\ \ \mu}$ is a curvature tensor of four-dimensional Riemann space of mRNA nucleotides. Vector of amino acid ζ^{μ} join mRNA triplet $\xi_k\xi_e\xi_n$, that is, vector ζ^{μ} depends only formally on a coding mRNA triplet. Therefore, the function $\Lambda_{ken}^{\ \ \mu}$ is a quantized presentation of the curvature tensor $R^{ken}_{\ \ \mu}$. Curvature tensor $R^{ken}_{\ \ \mu}$ of four-dimensional Riemann space has 20 independent components; accordingly, its quantized presentation $\Lambda_{ken}^{\ \ \mu}$ also has 20 independent vector presentations, that is, amino acids.

Degeneracy of the genetic code is determined by two causes:

1) formal relation between amino acid and coding mRNA triplet, that is, simple superposition of an amino acid on its mRNA triplet;
2) limited number of independent components of curvature tensor of four-dimensional Riemann space.

Note that curvature tensor of six-dimensional Riemann space has 105 independent components, which would be quite sufficient for non-degenerate correspondence of amino acids to mRNA triplets. However, in this case an amino acid would depend not only on mRNA triplet, but also on tRNA triplet, which would greatly complicate coupling between mRNA and tRNA.

We will show that, despite irreversibility of mRNA triplet transformation into amino acid, the volume of functional space remains unchanged at transition from coding of mRNA information to amino acid coding.

At transition from functions $\Omega_{t\beta}^{\ ken}$ to functions $\Xi_{\mu t\beta}$, functional volume is equal to $3^5 = 243$. To this volume there should be added the volume of auxiliary spaces, equal to $13 = 5 + 4 + 3 + 1$. Accordingly, we get 256 functions of mRNA triplet transformation into amino acids of the genetic code.

Reverse transition from functions $\Xi_{\mu t\beta}$ to functions $\Omega_{t\beta}^{\ ken}$ and from amino acids ζ^{μ} to triplet $\xi_k\xi_e\xi_n$ needs $5^3 + 3^1 = 128$ functions. In addition, 128 triplets of mRNA-tRNA pairing should be added to this number.

Thus, genetic code degeneration is connected with limited number of independent components of four-dimensional Riemann space of mRNA. Transformation of mRNA triplets into amino acids of the genetic code preserves the volume of the informational space.

Selected Bibliography

Buchatsky, L.P., Stcherbic, V.V. Degeneracy of the genetic code in four-dimensional Riemann space of matrix RNA. Bulletin of Kyiv National University 53: 71–72, 2008 (in Ukrainian).

Stcherbic, V.V., Buchatsky, L.P. Hidden symmetry of the genetic code. Problems of ecological and medical genetics and clinical Immunology 113: 132–142, 2012 (in Russian).

Stcherbic, V.V., Buchatsky, L.P. The tRNA molecule is genetic curvature tensor in four-dimensional affine space. Scientific Transactions of Ternopil National Pedagogical University. Ser. Biol. 54: 30–34, 2013 (in Ukrainian).

Structure of mRNA

Saenger, W. Principles of Nucleic Acid Structure. Berlin: Springer-Verlag, 1984.

Singer, M., Berg, P. Genes and genomes. California: University Science Books, 1991.

Watson, J.D., Baker, T.A., Bell, S.P., Gann, A., Levine, M., Losick, R. Molecular biology of the gene. New York: Cold Spring Harbor Laboratory, 2004.

Clark, D.P. Molecular biology. Understanding the genetic revolution. Elsevier Academic Press, 2005.

Weaver, R.F. Molecular biology. New York: McGraw-Hill, 2012.

Shi, H., Moore, P.B. The crystal structure of yeast phenylalanine tRNA at 1.93 Å resolution: A classic structure revisited. RNA 6: 1091–1105, 2000.

Aminoacyl-tRNA synthetases

Rich, A., Schimmel, P.R. Structural organization of complexes of transfer RNAs with aminoacyl transfer RNA synthetases. Nucl. Acids Res. 4: 1649–1665, 1977.

Cavarelli, J., Moras, D. Recognition of tRNAs by aminoacyl-tRNA synthetases. FASEB 7: 79–86, 1993.

Woese, C.R., Olsen, G.J., Ibba, M., Soll, D. Aminoacyl-tRNA synthetases, the genetic code, and the evolutory process. Microbiol. and Mol. Biol. Rev. 64: 202–236, 2000.

O'Donoghue, P., Luthey-Schulten, Z. On the evolution of structure in aminoacyl-tRNA synthetases. Microbiol. and Mol. Biol. Rev. 67: 550–573, 2003.

Ibba, M., Francklyn, C., Cusack, S. The aminoacyl-tRNA synthetases. Georgtown: Landes Bioscience, 2005.

Graifer, D.M., Moore, N.A. Protein biosynthesis. Novosibirsk: Novosibirsk University Press, 2011 (in Russian).

Co-evolution theory

Tze-Fei Wong, J. A Co-evolution theory of the origin of the genetic code. Proc. Nat. Acad. Sci. USA 72: 1909–1912, 1975.

Tze-Fei Wong, J. Coevolution theory of the genetic code at age thirty. BioEssays 27: 416–425, 2005.

Di Giulio, M. The coevolution theory of the origin of the genetic code. Physics Life Rev. 1: 128–137, 2004.

Di Giulio, M. An extension of the coevolution theory of the origin of the genetic code. Biol. Direct. 3: 37, 2008.

Origin and symmetry of the genetic code

Di Giulio, M. The origin of the genetic code: theories and their relationships, a review. BioSystems 80: 175–184, 2005.

Yarus, M., Caporaso, J.G., Knight, R. Origins of the genetic code: The escaped triplet theory. Annu. Rev. Biochem. 74: 179–198, 2005.

Hornos, J.E.M. and Hornos, Y.M.M. Algebraic model for the evolution of the genetic code. Phys. Rev. Lett. 71: 4401–4404, 1993.

Bashford, J.D., Tsohantjis, I., Jarvis, P.D. A supersymmetric model for the evolution of the genetic code. Proc. Natl. Acad. Sci. USA 95: 987–992, 1998.

Hornos, J.E.M., Hornos, Y.M.M., Forger, M. Symmetry and symmetry breaking: an algebraic approach to the genetic code. Int. J. Mod. Physics 13B: 2795–2885, 1999.

Antoneli, F., Forger, M., Gaviria, P.A. On amino acid and codon assignment in algebraic models for the genetic code. Int. J. Mod. Physics 24B: 435–463, 2010.

Négadi, T. Symmetry Groups for the Rumer–Konopel'chenko– Shcherbak "Bisections" of the genetic code and applications. Internet Electron. J. Mol. Des. 3: 247–270, 2004.

Forger, M., Sachse, S. Lie Superalgebras and the multiplet structure of the genetic code I: codon representations. J. Math. Phys. 41: 5407–5422, 2000.

Forger, M., Sachse, S. Lie Superalgebras and the multiplet structure of the genetic code II: branching rules. J. Math. Phys. 41: 5423–5444, 2000.

Karasev, V.A., Stefanov, V.E. Topological nature of the genetic code. J. Theor. Biol. 209: 303–317, 2001.

Sánchez, R., Morgado, E., Grau, R. The Genetic Code Boolean Lattice. MATCH Commun. Math. Comput. Chem. 52: 29–46, 2004.

Castro-Chavez, F. A tetrahedral representation of the genetic code emphasizing aspects of symmetry. BIO-Complexity 2: 1–6, 2012.

Yang, C.M. The naturally designed spherical symmetry in the genetic code. arXiv.org/abs/q-bio.BM/0309014, 2003.

Jimenez-Montano, M.A., de la Mora-Basanez, R., Poschel, T. The hypercube structure of the genetic code explains conservative and non-conservative amino acid substitutions *in vivo* and *in vitro*. BioSystems 39: 117–125, 1996.

Tlusty, T. A model for the emergence of the genetic code as a transition in a noisy information channel. J. Theor. Biol. 249: 331–342, 2007.

Tlusty, T. A relation between the multiplicity of the second eigenvalue of a graph Laplacian, Courant's nodal line theorem and the substantial dimension of tight polyhedral surfaces. Elec. J. Linear Algebra 16: 315–324, 2007.

Frappat, L., Sciarrino, A., Sorba, P. A crystal base for the genetic code. Phys. Lett. 250A: 214, 1998.

Frappat, L., Sorba, P., Sciarrino, A. Quantum Groups and the Genetic Code. Theor. Math. Phys. 128: 856–869, 2001.

Chipens, G.I. The hidden symmetry of the genetic code and rules of amino acid interactions. Russian Journal of Bioorganic Chemistry 17: 1335–1346, 1991 (in Russian).

Riemann geometry

Cartan, E. Spaces of affine, projective and conformal connection. Kazan: Kazan University Press, 1962 (in Russian).

Witten, L. Invariants of general relativity and classification of spaces. Phys. Rev. 113: 357–362, 1959.

Rashevsky, P.K. Riemann geometry and tensor analysis. Moscow: URSS Press, 2003 (In Russian).

Landau, L.D., Lifshitz, E.M. The Classical Theory of Fields, 4th Edition. Butterworth-Heinenann, 1987.

Grassmann algebra and Clifford algebra

Berezin, F.A. The Method of Second Quantization. New York: Academic Press, 1966.

Faddeev, L.D., Slavnov, A.A. Gauge Fields: Introduction to Quantum Theory. Addison-Wesley Publishing Company, 1991.

Shirokov, D.S. Clifford algebras and spinors. Moscow: Steklov Institute of Mathematics, RAS, 2011 (In Russian).

3

Fine-Structure Constant in Presentation of Biomolecules

Fine-structure constant $\alpha = e^2/\hbar c \approx 1/137$ (e—electron charge, $\hbar$—Planck constant, c—speed of light) characterizes intensity of electromagnetic interaction of elementary particles. In quantum electrodynamics, electrically charged particles interact due to exchange of virtual photons. Fine-structure constant arises as a nondimensional parameter, which characterizes intensity of this interaction. Fine-structure constant is one of values, which physicists have been measuring with increasing accuracy for many decades. This constant determines energy levels of electrons in atoms exactly. The fine structure of these levels emerges due to electric attraction of electrons to the nucleus and electromagnetic interactions between electrons.

The 2010 CODATA recommended value of the reciprocal of the fine-structure constant is given by $1/\alpha = 137{,}035999074 \pm 0{,}000000044$.

Constant α was introduced into physics by Sommerfeld in 1916 in order to explain the fine structure of energy levels of hydrogen atom. Initially, fine-structure constant was defined as the ratio of velocity of electron on lowest Bohr orbit to velocity of light.

The origin of this constant and its physical meaning are still not disclosed. Physicists believe that it implies something of great importance about the surrounding world. A special feature of fine-structure constant, namely, invariance to selection of unit system (abstractness of constant α), allows us to consider it the first candidate for the role of truly fundamental constant. A large number of researchers have been trying to understand the physical meaning of this constant, or to express it in a compact mathematical form.

Numerical value of reciprocal fine-structure constant, especially its integer part 137, is the subject of investigation of mathematicians who have been trying to find its group-theoretical basis.

A new approach in understanding physical meaning of the constant was proposed by Novoselov and Geim (2008). In their research, fine-structure constant defines optical transparency of hexagonal lattice monolayer of carbon atoms (graphene) in visible light rays. It became clear for the first time that the value of fine-structure constant might not only be applied to quantum electrodynamics but also may define physical properties of crystals.

FINE-STRUCTURE CONSTANT AND PROTEIN STRUCTURE

Our representation of fine-structure constant by amino acids of the genetic code is based on dependence of reciprocal constant $1/\alpha$ on the value of protonic charges of amino acid residues. We underline at once that protonic charge of amino acid residue (radical) does not coincide with the number of its protons.

The structure of amino acids of the genetic code

Each of the twenty amino acids of the genetic code consists of amino acid backbone **$CH(NH_2)COOH$** and amino acid residue (radical). Each radical will be defined by structure-vectorial linear field, which characterises the number of atoms: carbon **C**, hydrogen **H**, oxygen **O**, nitrogen **N** and sulphur **S**. Main characteristic of the radical structure is the number of protons or protonic charge of the radical. Let us assume that amino acid radical is electrically neutral, and with protonic charge, we will always associate corresponding number of electrons.

Let us define atoms **C, H, O, N, S** as basic vectors of protonic charges: $Q_p(\mathbf{C}) = 6$, $Q_p(\mathbf{H}) = 1$, $Q_p(\mathbf{O}) = 8$, $Q_p(\mathbf{N}) = 7$, $Q_p(\mathbf{S}) = 16$. Protonic charge of amino acid residue will be calculated through summation of protonic charges of atoms. Proline is the exception to basic structure of amino acids; therefore, its virtual amino acid residue may be determined after opening the pyrrolidine ring. This operation needs compensation. Among other amino acids, only tryptophan contains pyrrolidine ring, so it fits for changing its structure. In virtual benzene ring of tryptophan, **CH** group should be replaced by **CH_2** group. This will increase protonic charge of tryptophan by four units. Such changes result in presentation of genetic code triplets as elements of Clifford algebra $Cl(4, 2)$ with signature [36+ 28–]. Compensation of pyrrolidine ring disconnecting ($Q_p = -2$) is performed by the marked amino acid **His** (Commentary 3.1).

Table 3.1 shows characteristic of internal space of genetic code amino acids and the number of total protonic charges of amino acids and amino acid residues.

Amino acid characteristic is determined by the formula:

$$\rho(\text{Am}) = \begin{cases} +1, \text{ if } Q_p(\text{Am}) \; mod \; 8 = 0 \\ -1, \text{ otherwise} \end{cases}$$

Clifford algebra of amino acid radicals is defined on the basis of protonic charge Q_p(Amr) *mod* 4 and is equal to $Cl(7, 13) \cong Cl(14, 6)$. Pairs of amino acid residues generate Clifford algebra equal to $Cl(14, 6) \cong Cl(10, 10) \cong Cl(11, 9)$ on the basis of protonic charge Q_p(Amr) *mod* 8. Clifford algebra of amino acid residues pairs is isomorphic to Clifford algebra of free radicals of amino acids. Clifford algebra of free amino acids is determined on the basis of total protonic charge of amino acids according to the characteristic ρ(Am) and is equal to $Cl(12, 8) \cong Cl(9, 11)$.

In chapter 1, there was considered Clifford algebra $Cl(4, 2)$, in which triplet basis from twenty elements e_{abc} may be introduced with signature that also have value [12+ 8–].

Table 3.1 Characteristic of the internal space of amino acids.

Amino acid Am		Chemical formula of radical Amr	Protonic charge of radical Q_p(Amr)	Q_p(Amr) *mod* 4, *mod* 8	Protonic charge Q_p(Am)	ρ(Am)
A	Alanine	1C 3H	9	+1 +1	48	+1
C	Cysteine	1C 3H 1S	25	+1 +1	64	+1
D	Asparag. acid	2C 3H 2O	31	–1 –1	70	–1
E	Glutam. acid	3C 5H 2O	39	–1 –1	78	–1
F	Phenylalanine	7C 7H	49	+1 +1	88	+1
G	Glycine	1H	1	+1 +1	40	+1
H	Histidine	4C 5H 2N	43	–1 +3	82	–1
I	Isoleucine	4C 9H	33	+1 +1	72	+1
K	Lysine	4C 10H 1N	41	+1 +1	80	+1
L	Leucine	4C 9H	33	+1 +1	72	+1
M	Methionine	3C 7H 1S	41	+1 +1	80	+1
N	Asparagine	2C 4H 1O 1N	31	–1 –1	70	–1
P	Proline	3C 5H	23	–1 –1	62	–1
Q	Glutamine	3C 6H 1O 1N	39	–1 –1	78	–1
R	Arginine	4C 10H 3N	55	–1 –1	94	–1
S	Serine	1C 3H 1O	17	+1 +1	56	+1
T	Threonine	2C 5H 1O	25	+1 +1	64	+1
V	Valine	3C 7H	25	+1 +1	64	+1
W	Tryptophan	9C 12H 1N	73	+1 +1	108	–1
Y	Tyrosine	7C 7H 1O	57	+1 +1	96	+1

Total protonic charges of amino acid radicals may be grouped into 14 invariant protonic charges: 1, 9, 17, 23, 25, 31, 33, 39, 41, 43, 49, 55, 57, 73 (all numbers of protonic charges of amino acid radicals are odd).

To interpret reciprocal fine structure constant, we will need the structure of polypeptide chain. Periodic basis of polypeptide chain backbone, linked to amino acid residues, has protonic charge equal to 29. From the atoms **C**, **H**, **O**, **N** in the backbone structure, there may be 12 bases formed, with the number of hydrogen atoms less then six.

Periodic bases have the following structure:

- self-conjugated pair of direct basis $2Q_2$

  Q_2 = 2C 2H 1O 1N;
  Q_2 = 2C 2H 1O 1N.

  first conjugated pair of the basis $2Q_2$

  Q_1 = 1C 1H 1O 2N;
  Q_3 = 3C 3H 1O.

  second conjugated pair of the basis $2Q_2$

  Q_0 = 1C 2O 1N;
  Q_4 = 3C 4H 1N.

  third conjugated pair of the basis $2Q_2$

  $Q_{1,1}$ = 2C 1H 2O;
  $Q_{3,1}$ = 2C 3H 2N.

  Bases Q_2, Q_1, Q_3, Q_0, Q_4, $Q_{1,1}$, $Q_{3,1}$ may be skew or sector.

- first conjugated pair of the basis $2Q_1$

  $Q_{0,1}$ = 1O 3N;
  Q_2 = 2C 2H 1O 1N.

  second conjugated pair of the basis $2Q_1$

  $Q_{1,1}$ = 2C 1H 2O;
  $Q_{1,2}$ = 1H 4N.

  third conjugated pair of the basis $2Q_1$

  Q_0 = 1C 2O 1N;
  $Q_{2,1}$ = 1C 2H 3N.

- conjugated pair of the basis $2Q_{2,1}$

  $Q_{1,2}$ = 1H 4N;
  $Q_{3,1}$ = 2C 3H 2N.

- first conjugated pair of the basis $2Q_{3,1}$

  $Q_{1,2}$ = 1H 4N;
  Q_5 = 4C 5H.

  second conjugated pair of the basis $2Q_{3,1}$

  $Q_{2,1}$ = 1C 2H 3N;
  Q_4 = 3C 4H 1N.

- conjugated pair of the basis $2Q_3$

  $Q_{1,1}$ = 2C 1H 2O;
  Q_5 = 4C 5H.

- conjugated pair of the basis $2Q_4$

  $Q_{3,1}$ = 2C 3H 2N;
  Q_5 = 4C 5H.

Bases $Q_{0,1}$, $Q_{1,2}$, $Q_{2,1}$, Q_5, $Q_{5,1}$ have complex geometric configuration.

Basis, which is not included in the periodic basic configurations:

$Q_{5,1}$ = 5H 3O.

Bases, which have no pairs of conjugated bases:

Q_0, $Q_{0,1}$, $Q_{1,1}$, $Q_{1,2}$, Q_5, $Q_{5,1}$.

First indices in symbols $Q_0 \ldots Q_5$ coincide with the number of hydrogen atoms in corresponding basis of amino acid backbone.

Figures 3.1a–d show 7 bases $2Q_2$ of polypeptide chain backbone. Figure 3.1e shows how basis $Q_{2,1}$ disorders connection of basis Q_3.

Clifford algebra of amino acid radicals, relative to the number of hydrogen atoms more then five, is equal to *Cl*(10, 10). Amino acid radicals with the number of hydrogen atoms more then five and the number of carbon atoms more then six lead to ambiguity of the basis formed of atoms **C, H, O, N, S.**

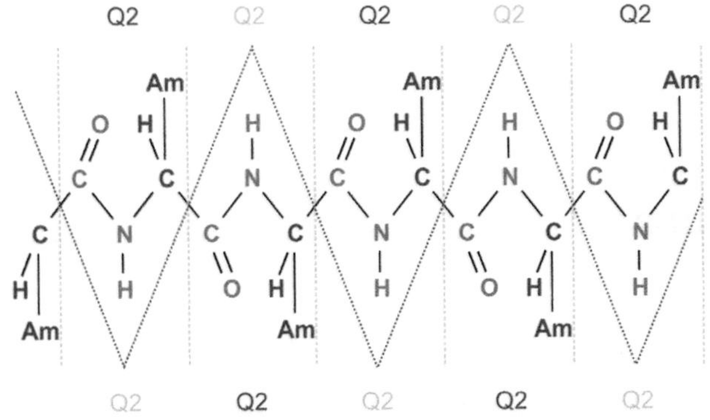

Figure 3.1a The structure of polypeptide chain with highlighted periodic bases of the backbone Q_2. Am are radicals of amino acids, which are connected with alternating sign to the atom C_α. The chain **CONH** is in the plane of peptide bond.

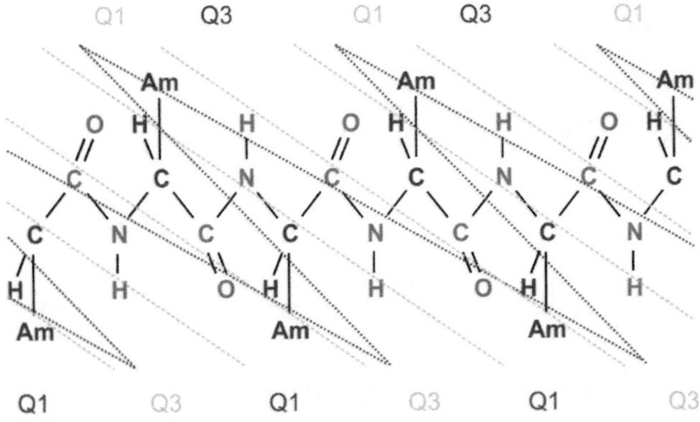

Figure 3.1b The structure of polypeptide chain with highlighted periodic bases of the backbone Q_1 and Q_3.

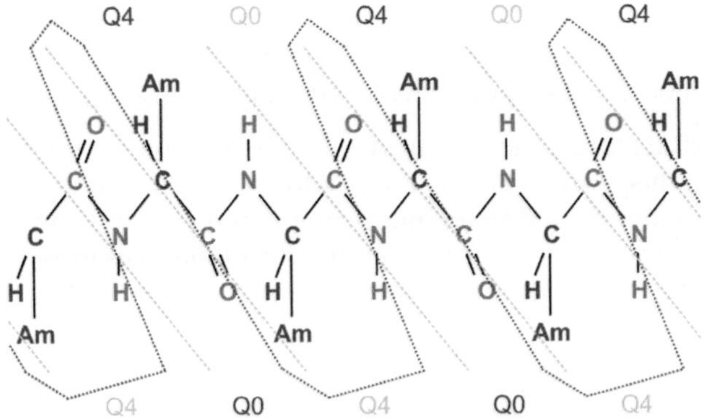

Figure 3.1c The structure of polypeptide chain with highlighted periodic bases of the backbone Q_0 and Q_4.

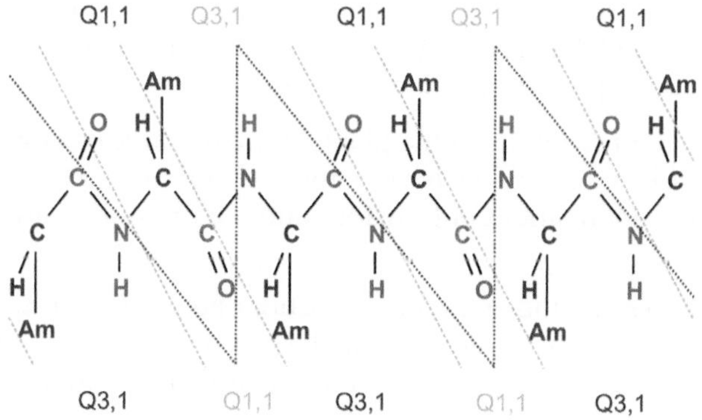

Figure 3.1d The structure of polypeptide chain with highlighted periodic bases of the backbone $Q_{1,1}$ and $Q_{3,1}$.

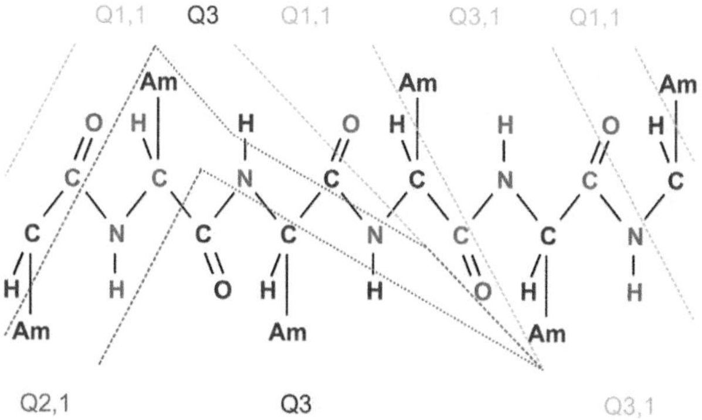

Figure 3.1e Disorder of connection of base Q_3 by base $Q_{2,1}$ with complex geometric configuration.

Interpretation of reversed fine-structure constant

Let us present reversed fine-structure constant as the sum of two parts: $1/\alpha = \alpha_0 + \delta\alpha_0$, where

$\alpha_0 = 137$ is the integer part of $1/\alpha$,
$\delta\alpha_0 = 0.035999074$ is the fractional part of $1/\alpha$.

Further, we will individually interpret integer and fractional parts of $1/\alpha$.

Let us decompose α_0 in three components

$$\alpha_0 = P(4) + \pi(6) + P(8),$$

where P(4) = 21, P(8) = 73 are the number of points of finite projective planes of 4th and 8th orders; $\pi(6) = 43$ is the number of fictitious points of nonexistent finite projective plane of 6th order.

We will identify points of the plane P(4) with radicals of genetic code amino acids together with the terminator. The value $\alpha_0 - P(4) = 116$ will be determined as an invariant of transformation of protonic charge of amino acid radicals.

Figure 3.2 shows the graph of connections between radicals of genetic code amino acids with the terminator. The function r^{-1}_1 determines reverse transfer of generating root $\varphi(1) = 1$ of terminator $\Theta(0)$ to glycine **G**(1).

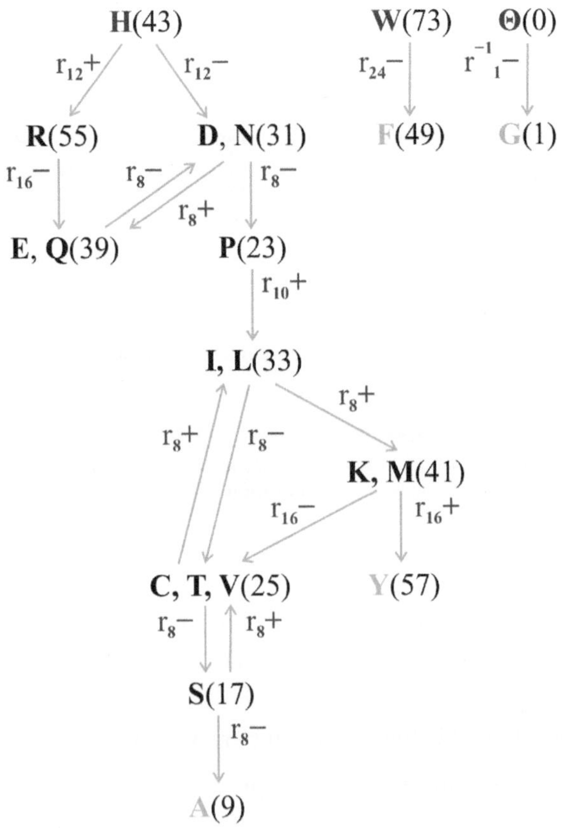

Figure 3.2 Graph of links between radicals of genetic code amino acids. rg± is the number of primitive roots (proper or improper) of protonic charge of amino acid radical: $g = \varphi(\varphi(Qp(Amr))$; φ is Euler numerical function (Commentary 3.2). Sign + or – corresponds to increase or decrease of protonic charge of amino acid radical at transferring of generating roots. Arrow shows the direction of generating root transferring. $\Theta(0)$ is the terminator with protonic charge equal to zero.

Invariance of the value $\alpha_0 - \mathrm{P}(4)$ determines identity

$$\mathbf{W}(73) + \mathbf{H}(43) + \Theta(0) = \mathbf{A}(9) + \mathbf{Y}(57) + \mathbf{F}(49) + \mathbf{G}(1).$$

Expressing amino acid residues through structure-vectorial field, we will have:

$$13\mathbf{C}\ 17\mathbf{H}\ 3\mathbf{N} + \Theta(0) = 15\mathbf{C}\ 17\mathbf{H}\ 1\mathbf{O} + \mathbf{G}(1),$$

and final identity $^{\Theta(0)}$:

$$3\mathbf{N} + \Theta(0) = 2\mathbf{C} + 1\mathbf{O} + \mathbf{G}(1),$$

where functions of terminator and glycine are highlighted.

It is clear that the number of protons (protonic charge) is saved in the identity $^{\Theta(0)}$. Let us suggest that baryonic charge is also saved. After selecting the neutron components in atom nuclei **N, C, O, G**(1) we will have the equation for the function $\Theta(n)$:

$$n + \Theta(n) = n(\mathbf{G}),$$

where $n(\mathbf{G}) = 0, 1, 2$ is the number of neutrons in hydrogen atom isotope.

Terminator function $\Theta(n)$ has three numerical values:

$$\Theta(n) = \begin{cases} \text{antineutron, if } n(\mathbf{G}) = 0; \\ 0 \text{ (no mRNA triplet), if } n(\mathbf{G}) = 1; \\ \text{neutron, if } n(\mathbf{G}) = 2. \end{cases}$$

In consideration of rare abundance of deuterium and tritium, we may state that the terminator in most cases is equivalent to antineutron. This equivalence we understand in the following sense. Terminator image (disintegration of the system composed of small and large subparticles of ribosome and mRNA) is determined by the physical process of antineutron decaying in three particles: $\tilde{n} \rightarrow p^- + e^+ + \nu_e$.

If terminator is absent, that is, 20 amino acids and fictive point of terminator Θ_f are placed in the plane P(4), the transition $\mathbf{G}(1) \leftrightarrow \Theta(0)$ will give the new identity $^{\mathrm{G}(1)}$:

$$3\mathbf{N} + \mathbf{G}(1) = 2\mathbf{C} + 1\mathbf{O} + \Theta_f,$$

$$p + n + \mathbf{G}(1) = \Theta_f.$$

It follows from the identity $^{\mathrm{G}(1)}$ that three fictitious states of the terminator (free proton, bound neutron and bound proton) become united into one state. This state is equivalent to helium atom isotope. For hydrogen atom isotopes with $n(\mathbf{G}) = 0, 1, 2$, fictitious terminator is equivalent to helium atom isotopes $\mathbf{He}^3$, $\mathbf{He}^4$, $\mathbf{He}^5$. We understand this equivalence in a way the image of behaviour of the fictitious terminator is the behaviour of helium atom isotopes in the matter.

The coevolution theory for amino acids of the genetic code is well known. Shown in Figure 3.2, graph of bonds between amino acid radicals does not coincide with metabolic ways of amino acid transformation. Nevertheless, these abstract bonds show that amino acids passed a highly complex path of evolutionary transformations.

Further, we will examine the fractional part of $1/\alpha$. To interpret the value $\delta\alpha_0$, we should take notice of the fact that the value $1/29 = 0.0344$ is very close to the experimental measurements. Therefore, we may suppose that $\delta\alpha_0$ may be calculated on the basis of expansion into inverse power series of some protonic charge of periodic basis of polypeptide chain backbone.

General formula of $\delta\alpha_0$-expansion in basis Q_i is as follows:

$$\delta\alpha_0 = \Sigma_n -(-1)^n \lambda_n(Q_i) / q_i^n,$$

where λ_n are expansion coefficients;

q_i is protonic charge of expansion in basis Q_i;

$n = 1, 2\ldots\ n_{max}$ is power of expansion.

The number n_{max} is not always equal to the number of denoted atoms (**H, C, O, N**) in the basis Q_i and it is corrected with regard to the order type.

For periodic bases of polypeptide chain backbone, the charge q_i may be determined as protonic charge 29 with the exception of the number (not equal to zero) of denoted atoms, which enter into the composition of the backbone basis.

Assume that denoted atom of expansion into a series of the value $\delta\alpha_0$ is hydrogen atom **H**. Then $q_0 = 29$, $q_1 = 28$, $q_2 = 27$, $q_3 = 26$, $q_4 = 25$, $q_{1,1} = 28$, $q_{3,1} = -26$. For the charge $q_{3,1}$ in the basis $Q_{3,1}$(**H**), minus sign is chosen to realize the identity $^{\alpha 0}$:

$$q_0 + q_1 + q_2 + q_3 + q_4 + q_{1,1} + q_{3,1} = 137.$$

On the basis of the constant α experimental value, we may calculate coefficients of expansion in bases $Q_1\ldots Q_4$ and make sure that they are close to one.

We will calculate coefficients of expansion in bases $Q_1\ldots Q_4$ by denoted atoms **H, C, O, N**, assuming that only coefficients $\lambda_1(Q_i)$ are not equal to 1 (Table 3.2a–d). The sum of coefficients $\lambda_1(Q_i) \approx 5$. The number of coefficients $\lambda_2(Q_i)$ is equal to 5. Protonic charge of expansion in basis Q_i is underlined, if denoted atom is absent.

Pairs of expansion charges $q_4(\mathbf{O}) = 27$, $q_3(\mathbf{N}) = -28$ and $q_{3,1}(\mathbf{O}) = -28$, $q_{1,1}(\mathbf{N}) = 27$ mutually complement each other and are chosen to realize the identity $^{\alpha 0}$.

Table 3.2a–d shows that the following relation is performed with high accuracy:

$$\lambda_1(Q_0) + \lambda_1(Q_1) + \lambda_1(Q_2) + \lambda_1(Q_3) + \lambda_1(Q_4) + \lambda_1(Q_{1,1}) + \lambda_1(Q_{3,1}) = 5.$$

Table 3.2a Formula of $\delta\alpha_0$ (**H**) expansion in different bases of protein backbone.

Basis of protein backbone Q_i	**Coefficient $\lambda_1(Q_i)$**	**Formula of $\delta\alpha_0$ expansion**
$Q_0^{\,1}$(H)	1.043976	$\lambda_1(Q_0)/29$
Q_1(H)	1.007974	$\lambda_1(Q_1)/28$
Q_2(H)	1.009012	$\lambda_1(Q_2)/27 - 1/27^2$
Q_3(H)	0.972958	$\lambda_1(Q_3)/26 - 1/26^2 + 1/26^3$
Q_4(H)	0.938442	$\lambda_1(Q_4)/25 - 1/25^2 + 1/25^3 - 1/25^4$
$Q_{1,1}^{\,2}$(H)	1.043689	$\lambda_1(Q_{1,1})/28 - 1/28^2$
$Q_{3,1}$(H)	− 0.975917	$-\lambda_1(Q_{3,1})/26 - 1/26^2 - 1/26^3$
$\sum_i \lambda_1(Q_i) = 5.040134$		

$q_0(\mathbf{H}) = 29$, $q_1(\mathbf{H}) = 28$, $q_2(\mathbf{H}) = 27$, $q_3(\mathbf{H}) = 26$,
$q_4(\mathbf{H}) = 25$, $q_{1,1}(\mathbf{H}) = 28$, $q_{3,1}(\mathbf{H}) = -26$.

Table 3.2b Formula of $\delta\alpha_0$ (**C**) expansion in different bases of protein backbone.

Basis of protein backbone Q_i	**Coefficient $\lambda_1(Q_i)$**	**Formula of $\delta\alpha_0$ expansion**
$Q_0^{\,2}$(C)	1.043689	$\lambda_1(Q_0)/28 - 1/28^2$
Q_1(C)	1.007974	$\lambda_1(Q_1)/28$
$Q_2^{\,1}$(C)	0.971976	$\lambda_1(Q_2)/27$
Q_3(C)	0.972958	$\lambda_1(Q_3)/26 - 1/26^2 + 1/26^3$
Q_4(C)	− 0.975917	$-\lambda_1(Q_4)/26 - 1/26^2 - 1/26^3$
$Q_{1,1}^{\,2}$(C)	1.009012	$\lambda_1(Q_{1,1})/27 - 1/27^2$
$Q_{3,1}^{\,3}$(C)	1.007640	$\lambda_1(Q_{3,1})/27 - 1/27^2 + 1/27^3$
$\sum_i \lambda_1(Q_i) = 5.037332$		

$q_0(\mathbf{C}) = 28$, $q_1(\mathbf{C}) = 28$, $q_2(\mathbf{C}) = 27$, $q_3(\mathbf{C}) = 26$,
$q_4(\mathbf{C}) = -26$, $q_{1,1}(\mathbf{C}) = 27$, $q_{3,1}(\mathbf{C}) = 27$.

Table 3.2c Formula of $\delta\alpha_0$ (**O**) expansion in different bases of protein backbone.

Basis of protein backbone Q_i	**Coefficient $\lambda_1(Q_i)$**	**Formula of $\delta\alpha_0$ expansion**
Q_0(**O**)	0.971976	$\lambda_1(Q_0)/27$
Q_1(**O**)	1.007974	$\lambda_1(Q_1)/28$
Q_2^2(**O**)	1.043689	$\lambda_1(Q_2)/28 - 1/28^2$
Q_3^3(**O**)	1.042414	$\lambda_1(Q_3)/28 - 1/28^2 + 1/28^3$
Q_4^2(**O**)	1.009012	$\lambda_1(Q_4)/27 - 1/27^2$
$Q_{1,1}^3$(**O**)	1.007640	$\lambda_1(Q_{1,1})/27 - 1/27^2 + 1/27^3$
$Q_{3,1}^2$(**O**)	- 1.043689	$-\lambda_1(Q_{3,1})/28 - 1/28^2$
$\sum_i \lambda_1(Q_i) = 5.039016$		

$q_0(\mathbf{O}) = 27$, $q_1(\mathbf{O}) = 28$, $q_2(\mathbf{O}) = 28$, $q_3(\mathbf{O}) = 28$,
$q_4(\mathbf{O}) = 27$, $q_{1,1}(\mathbf{O}) = 27$, $q_{3,1}(\mathbf{O}) = -28$.

Table 3.2d Formula of $\delta\alpha_0$ (**N**) expansion in different bases of protein backbone.

Basis of protein backbone Q_i	**Coefficient $\lambda_1(Q_i)$**	**Formula of $\delta\alpha_0$ expansion**
Q_0(**N**)	1.007974	$\lambda_1(Q_0)/28$
Q_1^1(**N**)	0.971976	$\lambda_1(Q_1)/27$
Q_2^2(**N**)	1.043689	$\lambda_1(Q_2)/28 - 1/28^2$
Q_3^3(**N**)	- 1.044965	$-\lambda_1(Q_3)/28 - 1/28^2 - 1/28^3$
Q_4^3(**N**)	1.042414	$\lambda_1(Q_4)/28 - 1/28^2 + 1/28^3$
$Q_{1,1}^2$(**N**)	1.009012	$\lambda_1(Q_{1,1})/27 - 1/27^2$
$Q_{3,1}^3$(**N**)	1.007640	$\lambda_1(Q_{3,1})/27 - 1/27^2 + 1/27^3$
$\sum_i \lambda_1(Q_i) = 5.037740$		

$q_0(\mathbf{N}) = 28$, $q_1(\mathbf{N}) = 27$, $q_2(\mathbf{H}) = 28$, $q_3(\mathbf{N}) = -28$,
$q_4(\mathbf{N}) = 28$, $q_{1,1}(\mathbf{N}) = 27$, $q_{3,1}(\mathbf{N}) = 27$.

While calculating $\delta\alpha_0(\mathbf{H})$, coefficient $\lambda_1(Q_0)$ might be supposed to be equal to zero, since hydrogen atoms are absent in the basis Q_0. Then $\sum_i \lambda_1(Q_i) = 3.996158$. But this value of sum of coefficients $\lambda_1(Q_i)$ is not invariant at transition to bases of fractional part $\delta\alpha_0$ expansion by content of atoms **C, O, N**.

If $\sum_i \lambda_1(Q_i)$ is no invariant of expansion, only two expansions $\delta\alpha_0(\mathbf{H})$ and $\delta\alpha_0(\mathbf{C})$ will represent the integer part of $1/\alpha$. Thus, numerical value of the constant $1/\alpha$ is divided into two parts, integer and fractional.

Integer part of the constant $1/\alpha$ is presented by the number of amino acids (20), as well as by amino acid residues of the six amino acids—*histidine, tryptophan, alanine, glycine, phenylalanine* and *tyrosine*—that may be generating spinors of four-dimensional curvature tensor in the affine space (Chapter 2). Fractional part of the reciprocal fine-structure constant $1/\alpha$ is determined only by the polypeptide chain backbone and is presented by inverse power series of some protonic charge in each of seven periodic bases of protein backbone.

FINE-STRUCTURE CONSTANT AND ATP MOLECULE

Adenosine triphosphate (ATP) consists of adenosine, including adenine and ribose sugar, and three phosphate groups. ATP is a universal source of energy for all biochemical processes taking place in living organisms.

ATP is a chemical substance containing high-energy bonds, the hydrolysis of which leads to release of considerable quantity of energy.

Hydrolysis of macroergic bonds of adenosine triphosphate molecule results in removal of the farthest phosphate residue with formation of adenosine diphosphate (ADP) or in removal of two final phosphate residues with formation of adenosine monophosphate (AMP):

$$\text{ATP} + \mathbf{H_2O} \rightarrow \text{ADP} + \mathbf{H_3PO_4} + \text{energy (30.5 kJ/mole);}$$
$$\text{ATP} + \mathbf{H_2O} \rightarrow \text{AMP} + \mathbf{H_4P_2O_7} + \text{energy (45.6 kJ/mole).}$$

This reaction is the basic chemical reaction, which supplies numerous biochemical processes with energy.

ATP is also the starting substance at synthesis of adenine and the coding nucleotide in content of RNA and DNA.

Renewal of ATP molecules takes place on mitochondrial membranes by means of ADP phosphorylation:

$$\text{ADP} + \mathbf{H_3PO_4} + \text{energy} \rightarrow \text{ATP} + \mathbf{H_2O}.$$

Decomposition of one glucose molecule results in synthesis of 38 ATP molecules.

E.M. Galimov (2001) considers ATP as a key molecule (molecule No. 1) on the path of life evolution. This is due to the fact that reaction ATP → ADP is easily coupled with synthesis of biopolymers from monomers, when formation of covalent bonds between them are accompanied by water removal (polysaccharides, proteins). Besides, adenine may be easily synthesised abiogenously, that is, without participation of living organisms.

Adenosine triphosphate does not possess catalytic properties, but in the path of living matter evolution it became the first element of genetic information with its own store of free energy.

Information has the abstract numerical form, non-connected with specific physical quantities—energy, entropy and physical carrier of information.

Henry Quastler (1964) affirmed that

"The "accidental choice remembered" is a mechanism of creating information".

How random is selection of adenosine triphosphate among other sources of energy? Whereby is it more preferable than, for instance, guanosine triphosphate?

Priority of ATP is caused by its connection with an abstract number—fine-structure constant. ATP molecule integrates most important properties of living matter evolution:

- primary energy—sugar;
- storage of energy—triphosphate;
- primary genetic element—adenine;
- numerical expression of genetic information—connection with fine-structure constant;
- prototype of complementary pairs: adenine—ribose.

To substantiate these statements, we will consider the structure of ATP in detail (Figure 3.3).

Let us designate: 3-phosphate = $<\mathbf{PPP}>$, $\{\text{ribose} - \mathbf{CH_2}\} = \mathbf{R}ib$.

3-phosphate together with methylene part of ribose $\mathbf{CH_2}$ present integer part of the reciprocal fine-structure constant, since

$$Q_p\{<\mathbf{PPP}> + \mathbf{CH_2}\} = Q_p(1\mathbf{C}\ 6\mathbf{H}\ 10\mathbf{O}\ 3\mathbf{P}) = \alpha_0 = 137.$$

It is no accident that nucleotide adenine is the first genetic informational element, since $\{\text{3-phosphate} + \mathbf{CH_2}\}$ is a skew reflection of hydrogen atom **H** relative to adenine **A**:

$$Q_p\{<\mathbf{PPP}> + \mathbf{CH_2}\} = -Q_p(\mathbf{H}) + 2Q_p(\mathbf{A}).$$

(Note that skew reflection of vector $\mathbf{r}_1$ relative to vector $\mathbf{r}_2$ is given by $\mathbf{r}_1' = -\mathbf{r}_1 + 2\mathbf{r}_2$).

To understand the role of ribose in ATP, note that

$$Q_p\{<\mathbf{PPP}> + \text{ribose} + \mathbf{A}\} = 260 \sim \{2^8\} = Cl(8).$$

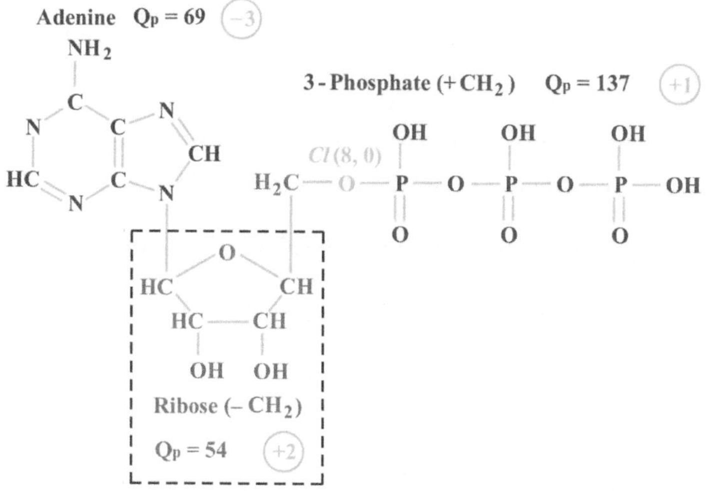

Figure 3.3 Structure of ATP. Ribose methylene group $\mathbf{CH_2}$ is joined to 3-phosphate. Highlighted oxygen atom **O** is a generator of Clifford algebra $Cl(8, 0)$. A number inside the circle is deviation of protonic charge of chemical structure from the structure of Clifford algebra.

Suppose that the pair adenine + ribose is a presentation of some Clifford algebra. To substantiate this assumption, let us assume that ATP is a presentation of Clifford algebra $Cl(8, 0)$ (so far, we will not consider isomorphic algebras $Cl(0, 8)$, $Cl(4, 4)$, $Cl(5, 3)$, $Cl(1, 7)$). Clifford algebra $Cl(8, 0)$ contains subalgebra of even elements, odd elements do not form any algebra. We know that Clifford algebra $Cl(p, q)$ contains Clifford subalgebra $Cl(q, p - 1)$. It follows from the relation $Cl(p, q) \cong Cl(q + 1, p - 1)$, which we will name **Clifford inversion.**

Clifford algebra $Cl(0, 7)$ = [72+ 56–] is subalgebra of algebra $Cl(8, 0)$. In ATP, the role of odd elements of Clifford algebra $Cl(8, 0)$ is fulfilled by 3-phosphate (+1), while the function of subalgebra $Cl(0, 7)$ is performed by the pair **A** + **R***ib*.

Suggest that [72+ 56–] = [**A(–3)**+ **R***ib***(+2)**–] is an even part of Clifford algebra $Cl(8, 0)$, then [<**PPP**> – **O**](**+1**) is an odd part of Clifford algebra $Cl(8, 0)$, and oxygen atom **O** is the charge-separator of these parts. Here in parentheses, there are shown protonic charges of the chemical structure deviation from structural parts of Clifford algebra.

Thus, adenine and ribose are positive and negative parts of Clifford algebra $Cl(0, 7)$. Algebra $Cl(7, 0)$ = [64+ 64–] has the appearance [**A(+5)**+ **ribose**(–2)–]; it annihilates **O**-charge-separator and is not a subalgebra of the algebra $Cl(8, 0)$. Instead of **O**-charge-separator, there may be considered $\mathbf{CH_2}$-charge-separator. Clifford algebra of $\mathbf{CH_2}$ group is equal to $Cl(6, 2)$ and contains subalgebra *Cl*(2, 5) = [56+ 72–], which has the appearance [**R***ib*(–2)+ **A**(+3)–]. But, in this case, balance of charges of deviation of ATP chemical structure from structural parts of Clifford algebra is equal to +2 (no ±1) relative to $\mathbf{CH_2}$-charge-separator, which prevents the correct interpretation of fractional part of inverse fine-structure constant $\delta\alpha_0$.

The pair **A** + **R***ib* is a preimage of complementary pairs **AT** and **GC**, since opposite "orientation" of **A** relative to **R***ib* may be determined in the following way. Let us select 5-ring **NCCNC(H)** in adenine, and 5-ring **CCCCO(0)**—in ribose. Each 5-ring is a basis of the pair **A** + **R***ib* with $Q_p = 32$ ($\mathbf{Z}_2$-grading of Clifford algebra *Cl*(6)); at that, 5-ring of ribose is an even part, that is, subalgebra *Cl*(5). Different parity of 5-rings is a preimage of different nucleotide orientation in complementary pairs. Negativeness of **R***ib* is determined by the even part of Clifford algebra *Cl*(6).

Connection of ATP with fractional part of the constant 1/α

Presence of **O**-charge-separator in ATP structure is connected with fractional part of the inverse fine-structure constant $\delta\alpha_0$. Balance of charges of deviation of ATP structure from structural parts of Clifford algebra is equal to **A**(–3) + **R***ib*(+2) = –1 for even part of Clifford algebra *Cl*(8, 0) with protonic charge Q_p{**A**(–3) + **R***ib*(+2)} = 123, and it equals to +1 for the structure {<**PPP**> – **O**} of the odd part of Clifford algebra *Cl*(8, 0) with protonic charge Q_p{<**PPP**> – **O**} = 129. Fractional part of 1/α is a presentation of Clifford inversion $(122_{-1} + 128_{+1})/8 \rightarrow 9/250 = \delta\alpha_0 = 0.036$ relative to **O**-charge-separator.

FINE-STRUCTURE CONSTANT AND DNA STRUCTURE

Adenine nucleotide **A**, the first genetic informational element, together with ribose ($-\mathbf{CH_2}$) generates Clifford algebra *Cl*(0, 7). In DNA, nucleotide **A** couples with nucleotide **T**. We will also show that in DNA **AT** pair generates Clifford algebra *Cl*(0, 7), while **GC** pair generates Clifford algebra *Cl*(6, 1). Protonic charge of **AT** and **GC** pairs coincide and is equal to 134 (Chapter 1). Separator of **AT** pair is Clifford algebra *Cl*(2, 1), while separator of **GC** pair is Clifford algebra *Cl*(1, 2). Both algebras are Clifford algebras *Cl*(3) with three generators (**Gen** *Cl* = 3).

Complementary DNA pairs are a presentation of integer part of fine-structure constant, since

$$Q_p\{\mathbf{A} + \mathbf{Gen}\ Cl(2, 1) + \mathbf{T}\} = Q_p\{\mathbf{G} + \mathbf{Gen}\ Cl(1, 2) + \mathbf{C}\} = \alpha_0 = 137.$$

Figure 3.4 shows how complementary DNA pairs together with chain backbone (deoxyribose and phosphate part) generate *doubled* presentation of constant $1/\alpha$.

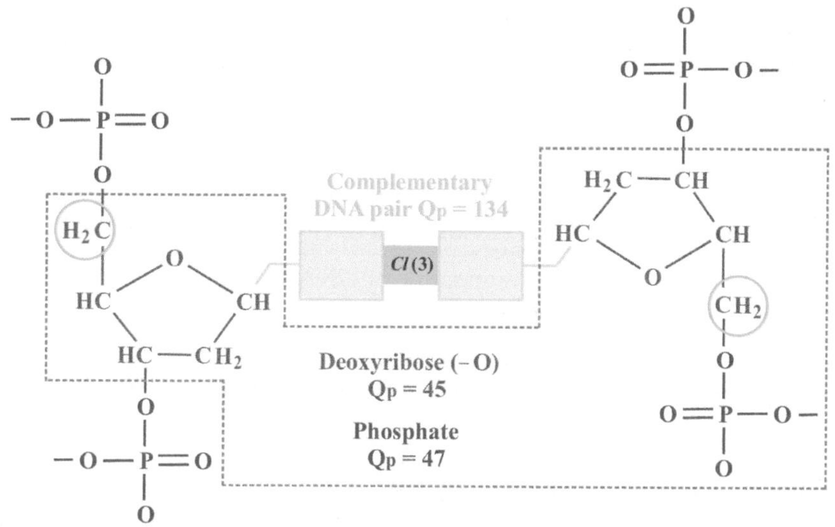

Figure 3.4 Doubled presentation of constant $1/\alpha$. Left $\mathbf{H_2C}$ and right $\mathbf{CH_2}$ charge-separators of two isomorphic Clifford algebras *Cl*(2, 6) and *Cl*(6, 2), accordingly, are highlighted in circles. Selection of phosphate—left or right—is conditional, but it should correspond to the nucleotide already connected to the DNA chain (not added in future).

Two residues of deoxyribose d**Rib**(−**O**) and one phosphate $<\mathbf{P}>$ are a presentation (twice) of integer part of the fine-structure constant since

$$Q_p\{\, 2d\mathbf{Rib}(-\mathbf{O}) + <\mathbf{P}>\} = \alpha_0 = 137.$$

Suppose that **AT** pair is a presentation of Clifford algebra *Cl*(0, 7) = [72+ 56−] = [**A**(−3)+ **T**(+9)−], where, with account taken of nucleotide orientation, protonic charges of deviation of nucleotide chemical structure from structural parts of Clifford algebra are shown in parentheses. Lorentz interval of deviation is equal to [1+ 3−].

Further, suppose that **GC** pair is a presentation of Clifford algebra *Cl*(6, 1) = [56+ 72−] = [**G**(+21)+ **C**(−15)−]. In both cases, the balance of the pair deviation is equal to $\delta Cl_{pair} = 6$. Parallel shift **AT** → **GC** is possible, since translated Clifford algebras *Cl*(21, 3) and *Cl*(9, 15) are isomorphous.

The variant, at which **AT** and **GC** pairs are presentations of Clifford algebras *Cl*(6, 1) and *Cl*(0, 7), respectively, is not possible owing to absence of translated Clifford algebra.

Four residues of deoxyribose d**R**ib(−**O**), two phosphates < **P** > and two charge-separators $\mathbf{H_2C}$ and $\mathbf{CH_2}$ are a dimeric presentation of Clifford algebra $Cl(8, 0) = Cl(2, 6) \oplus Cl(6, 2)^{\pi}$:

$$Cl(2, 6) \oplus Cl(6, 2)^{\pi} = [136+\ 120-] =$$
$$= [\{2\text{d}\mathbf{R}\text{ib}(-\mathbf{O}) + <\mathbf{P}>\}(+1) + \{2\text{d}\mathbf{R}\text{ib}(-\mathbf{O} - \mathbf{H_2C} - \mathbf{CH_2}) + <\mathbf{P}>\}(-1)-].$$

The balance of deviation of Clifford algebra $Cl(2, 6) \oplus Cl(6, 2)^{\pi}$ from doubled algebra $Cl(7, 0) \oplus Cl(7, 0)$ is equal to

$$\delta Cl(6, 2) = [128 - Q_p\{6+\ 2-\} - 121 = +3] + [128 - 137 = -9] = -6$$

with Lorentz interval of deviation [3+ 1−] with known properties of duplication (Chapter 1). It is obvious that $\delta Cl(6, 2)$ compensates δCl_{pair}.

Surprisingly that charge-separator of DNA nucleotides is not 3′(**O**)-, but 5′($\mathbf{CH_2}$)-end of deoxyribose.

Doubled fractional part of the constant $1/\alpha$ is equal to $2\delta\alpha_0 = 9/125$. This number may be determined, e.g., for **AT** and **TA** pairs, by the formula

$$2\delta\alpha_0 = [Q_p\mathbf{Gen}\ Cl(2, 1) + \delta Cl_{pair}]/[Q_p(\mathbf{AT}) - Q_p\mathbf{Gen}\ Cl(2, 1) - \delta Cl_{pair}],$$

while for DNA chain backbone—by the formula

$$-2\delta\alpha_0 = [128 - \alpha_0]/[121 + Q_p\{6+\ 2-\}].$$

Conclusion: fractional part of inverse fine-structure constant is completely compensated in complementary DNA pairs.

FINE-STRUCTURE CONSTANT AND CAP-STRUCTURE OF mRNA

Initiation of transcription results in modification of mRNA 5′-end, which is accompanied by joining of cap-structure. Capping is one of the very early modifications of growing RNA chains, which takes place after polymerisation of its first 25–30 nucleotides.

Capping is realized by three enzymes: RNA triphosphatase, guanylyl transferase and guanine-7-methyltransferase.

Cap is 7-methylguanosine joined by 5′,5′-triphosphate bridge with first nucleotide residue of initial mRNA. Modifications happen with first two nucleotides of mRNA: methylation by 2′-**O**-position of ribose.

Cap promotes splicing of pre-mRNA, export of mRNA from the nucleus, its effective translation and protection from fast degradation.

Figure 3.5 shows the structure of mRNA 5′-end.

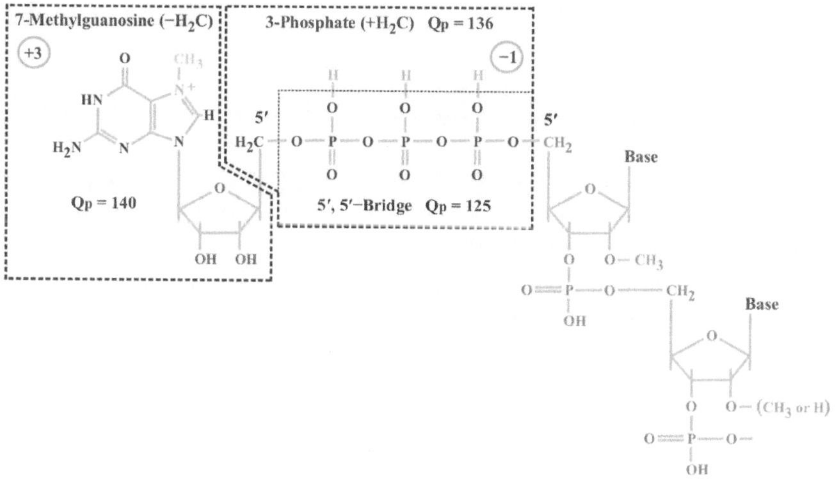

Figure 3.5 Structure of 5′-end of cap-mRNA. Metric of mRNA translation intervals [3+ 1–], joined to integer values of the constant $1/\alpha$, is highlighted in circles.

It was shown in Chapter 1 that excessive neutron of 7-methylguanine and three hydrogen atoms of methyl group $\mathbf{CH_3}$ generate metric of mRNA translation intervals: $[\mathbf{H}_3\ n]$ = [3+ 1–]. Cap-structure multiplies this metric first of all on doubled presentation of fine-structure constant and on ionic-inverse structure composed of $\mathbf{N^+}$ 7-methylguanine and $\mathbf{O^-O^-O^-}$ 5′,5′-triphosphate bridge.

Cap represents two copies of the inverse fine-structure constant by module of protonic charge 137: 7-methylguanosine ($-\mathbf{H_2C}$) with positive residue +3, and 3-phosphate ($+\mathbf{H_2C}$) with negative residue –1. These residues generate metrics of mRNA translation intervals: [3+ 1–].

The second copy of mRNA translation intervals [3+ 1–] is determined on the basis of electron donor $\mathbf{N^+}$ 7- methylguanine (–1) and proton donor 5′,5′-triphosphate bridge (+3), that is, inversely relative to the electric charge of these structures.

Doubled fractional part $2\delta\alpha_0$ is determined by ratio of protonic charge of methyl group of guanine (+9) to protonic charge of

5′,5′-triphosphate bridge (+125): $2\delta\alpha_0 = Q_p(\mathbf{CH_3})/Q_p(5',5'\text{-bridge})$.

APPENDIX 3.1

CLIFFORD ALGEBRAS OF SIMPLE CHEMICAL COMPOUNDS

Since $Cl(8, 0) \cong Cl(0, 8)$, the choice of Clifford algebra $Cl(8, 0)$ with positive part of the signature, equal to protonic charge of oxygen atom, may be supposed to be the conditional one.

On the whole, single atom is Clifford algebra $Cl(\theta, 0)$ or $Cl(0, \theta)$, where θ is protonic charge of the atom. For carbon atom **C**, two nonisomorphic algebra $Cl(6, 0)$ and $Cl(0, 6)$ are possible.

Compounds of two chemical elements is Clifford algebra of the type $Cl(\theta, \tau)$ or $Cl(\tau, \theta)$, where θ is protonic charge of one atom and τ—protonic charge of the other.

Clifford algebras $Cl(6, 2)$, $Cl(2, 6)$ of the compound $\mathbf{CH_2}$ are isomorphic.

We have isomorphism chains of Clifford algebras of chemical compounds:

$$Cl(\text{H}, \mathbf{N}) \sim Cl(\mathbf{O}, 0); \qquad Cl(\mathbf{N}, \text{H}) \sim Cl(\text{H}_2, \text{C});$$
$$Cl(\text{O}, \text{H}) \sim Cl(\text{H}_2, \mathbf{N}) \sim Cl(\text{C}, \text{H}_3); \qquad Cl(\text{H}, \mathbf{O}) \sim Cl(\mathbf{O}\text{H}, 0);$$
$$Cl(\text{N}, 0) \sim Cl(\text{H}, \text{C}) \sim Cl(\text{H}_3, \text{H}_4); \qquad Cl(\text{C}, 0) \sim Cl(\text{H}_2, \text{H}_4);$$
$$Cl(\text{S}, 0) \sim Cl(\text{CC}, \text{H}_4) \sim Cl(\mathbf{O}, \mathbf{O}) \sim Cl(\text{OH}, \mathbf{N}) \sim Cl(\text{H}, \mathbf{P});$$
$$Cl(\mathbf{O}, \text{H}_2) \sim Cl(\mathbf{N}, \text{H}_3) \sim Cl(\text{H}_4, \text{C}) \sim Cl(0, \text{H}_2\mathbf{O});$$
$$Cl(\text{H}_2, \mathbf{O}) \sim Cl(\text{C}, \text{H}_4) \sim Cl(\mathbf{O}\text{H}, \text{H}) \sim Cl(\text{H}_2\text{O}, 0) \sim Cl(\text{H}, \text{NH}_2);$$
$$Cl(\mathbf{P}, \text{OH}) \sim Cl(\text{H}_2\mathbf{O}, \text{CO}) \sim Cl(\text{NN}, \text{CH}_4) \sim Cl(\text{HH}_2\text{O}, \text{NC}) \sim$$
$$\sim Cl(\mathbf{N}, \mathbf{OOH}) \sim Cl(\text{H}_3, \mathbf{P}\text{C}) \sim Cl(\text{CCC}, \text{C}) \sim \text{Cl}(\text{SC}, \text{H}_2) \sim$$
$$\sim Cl(\text{C}, \text{PH}_3) \sim Cl(\text{H}_2, \mathbf{NCOH})$$

One may attempt to ascertain correspondence of a chemical formula to Clifford algebra signature by some indication of equivalence in both genetic and biological meaning.

If uracil **U** = 4**C** 3**H** 2**O** 2**N** is equivalent to cytosine **C** = 4**C** 4**H** 1**O** 3**N** (in third position of mRNA triplet of the genetic code), oxygen **O** will be equivalent to imino group **NH**. Then, $Cl(\mathbf{H}, \mathbf{N}) \cong Cl(\mathbf{O}, \mathbf{0})$, and choice of nitrogen **N** position will be well-defined.

If inosine **I** = 5**C** 3**H** 1**O** 4**N** is equivalent to adenine **A** = 5**C** 4**H** 5**N** (in first position of tRNA anticodon), oxygen **O** will be equivalent to imino group **NH** and hydroxile **OH** will be equivalent to amino group $\mathbf{NH_2}$. Then, $Cl(\mathbf{O}, \mathbf{H}) \cong Cl(\mathbf{H_2}, \mathbf{N})$, and choice of oxygen **O** and nitrogen **N** positions will be well-defined.

APPENDIX 3.2

ALGEBRAIC FORMULAE OF THE CONSTANT 1/α

Rosen formula

Let us designate with the letter ξ potentials of electromagnetic field. Commutator of electromagnetic fields in points x, y of four-dimensional space-time is as follows:

$$[\xi_\mu(x), \xi_\nu(y)] = 4\pi i(\hbar c/e^2)D(x-y)g_{\mu\nu},$$

where

$D(x-y)$ is photonic function of propagation;
$g_{\mu\nu}$ is metric tensor of Lorentz interval;
$\hbar c/e^2 = 1/\alpha$;
$\mu, \nu = 1\ldots4$.

Constant $4\pi/\alpha$ is equal to 1722.045 and distinguishes insignificantly from the integer value $1722 = 42 \times 41$.

Rosen (1976) proposed to divide each electromagnetic potential to parts numbered with elements of transitive group of the order 42:

$$[\xi_\mu{}^r(x), \xi_\nu{}^s(y)] = \begin{cases} iD(x-y)g_{\mu\nu}, & \text{if } r \neq s \\ 0, & \text{if } r = s \end{cases}$$

where $r, s = 1\ldots42$.

New commutator is equal to:

$$[\xi_\mu(x), \xi_\nu(y)] = \sum_{r,s=1\ldots42} [\xi_\mu{}^r(x), \xi_\nu{}^s(y)] = 42 \times 41 iD(x-y)g_{\mu\nu}$$

and determines rather well the value of the constant $1/\alpha \approx 137.0324$.

For complete description, we present below classes of conjugate elements of transitive group of the order 42:

1^7

```
(1) (2) (3) (4) (5) (6) (7)
```

1, 6

```
(1) (2, 6, 5, 7, 3, 4)
(1, 4, 5, 3, 7, 6) (2)
(1, 7, 2, 5, 6, 4) (3)
(1, 3, 6, 7, 5, 2) (4)
(1, 6, 3, 2, 4, 7) (5)
(1, 2, 7, 4, 3, 5) (6)
(1, 5, 4, 6, 2, 3) (7)
```

1, 6

```
(1) (2, 4, 3, 7, 5, 6)
(1, 6, 7, 3, 5, 4) (2)
(1, 4, 6, 5, 2, 7) (3)
(1, 2, 5, 7, 6, 3) (4)
(1, 7, 4, 2, 3, 6) (5)
(1, 5, 3, 4, 7, 2) (6)
(1, 3, 2, 6, 4, 5) (7)
```

$1, 3^2$

```
(1) (2, 3, 5) (4, 7, 6)
(1, 7, 5) (2) (3, 4, 6)
(1, 2, 6) (3) (4, 7, 5)
(1, 5, 6) (2, 7, 3) (4)
(1, 4, 3) (2, 6, 7) (5)
(1, 3, 7) (2, 5, 4) (6)
(1, 2, 4) (3, 6, 5) (7)
```

$1, 3^2$

```
(1) (2, 5, 3) (4, 6, 7)
(1, 5, 7) (2) (3, 6, 4)
(1, 6, 2) (3) (4, 5, 7)
(1, 6, 5) (2, 3, 7) (4)
(1, 3, 4) (2, 7, 6) (5)
(1, 7, 3) (2, 4, 5) (6)
(1, 4, 2) (3, 5, 6) (7)
```

$1, 2^3$

```
(1) (2, 7) (3, 6) (4, 5)
(1, 3) (2) (4, 7) (5, 6)
(1, 5) (2, 4) (3) (6, 7)
(1, 7) (2, 6) (3, 5) (4)
(1, 2) (3, 7) (4, 6) (5)
(1, 4) (2, 3) (5, 7) (6)
(1, 6) (2, 5) (3, 4) (7)
```

7

```
(1, 2, 3, 4, 5, 6, 7)
(1, 3, 5, 7, 2, 4, 6)
(1, 4, 7, 3, 6, 2, 5)
(1, 5, 2, 6, 3, 7, 4)
(1, 6, 4, 2, 7, 5, 3)
(1, 7, 6, 5, 4, 3, 2)
```

Wyler formula

Wyler formula (1969) is based on calculation of ratios of volumes of continuous groups associated with the conformal group SO(5, 2). Such physical reasons as, e.g., "geometrical possibility of electron to emit and absorb virtual photon defines value of fine-structure constant", are incorrect, since in quantum electrodynamics, there are no direct link with Feynman diagram for calculating probabilities of physical processes.

Wyler formula is a very precise formula for calculating the constant α. Just because of this, it is paid so much attention.

Wyler formula is:

$$\alpha = \frac{2V(S2)[V(D5)]^{\frac{1}{4}}}{V(S4)V(Q5)} = \frac{9}{8\pi^4}\left(\frac{\pi^5}{2^4 \cdot 5!}\right)^{\frac{1}{4}} = \frac{1}{137.03608}$$

Here

V(S2), V(S4) are "the area" of sphere surface in three-dimensional and five-dimensional spaces;

V(Q5) = V(S4 × RP1) is "the area" of five-dimensional disc boundary;
V(D5) is the volume of five-dimensional disc.

Numerical values of the parameters are:

$V(S2) = 4\pi$; $V(S4) = 8\pi^2/3$; $V(Q5) = 8\pi^3/3$; $V(D5) = \pi^5/(2^4 \times 5!)$;
$V(S2) \approx 12.5664$; $V(S4) \approx 26.3189$; $V(Q5) \approx 82.6834$; $V(D5) \approx 1/6.2741$.

Suppose that Q_m is protonic charge of generalized periodic basis of polypeptide chain backbone. Then, $V(S2) \sim Q_m/2$, $V(S4) \sim Q_m$, $V(Q5) \sim 3Q_m$, $V(D5) \sim 4/Q_m$. The equation $(4/Q_m^5)^{1/4} = 3\alpha$ gives $Q_m = 28.0661$. It follows from Wyler formula that generalized periodic basis of polypeptide chain backbone contains only one hydrogen atom.

Polydisc $D5 = SO(5, 2)/SO(5) \times SO(2)$ correlates with presentation of the identity $^{\alpha 0}$. More difficult is to understand the presence of fourth root from the polydisc volume. Formal explanation is that transition of noncompact group SO(5, 2) to compact group SO(5) × SO(2) is some imaging with the factor $[V(D5)]^{1/4}$.

Bases $Q_{0,1}$, $Q_{1,2}$, $Q_{2,1}$, Q_5, $Q_{5,1}$ of the polypeptide chain backbone give noncompact presentation of the value $\delta\alpha_0$(**H**). However, if they are substituted with bases Q_0, Q_1, $Q_{1,1}$, Q_2, Q_4, we will have compact presentation of the value $\delta\alpha_0$(**H**) with substitution factor 7/11, since at that bases Q_3, $Q_{3,1}$ will not be destroyed.

The value $[V(D5)]^{1/4} \approx 1/1.5827 \sim 7/11$ becomes plausible at changing noncompact basis SO(5, 2) with compact one SO(5) × SO(2).

Generally, conformal group SO(5, 2) contains "two dimensional time" and has no clear physical interpretation.

COMMENTARY 3.1

Compensation of pyrrolidine ring opening in radical of amino acid **Pro** is possible even without increasing protonic charge of amino acid **Trp** radical. For this, we will point out amino acid **His** with metric tensor of Lorentz interval and leave protonic charge of amino acid **Trp** unchanged: Q_p(9**C** 8**H** 1**N**) = 69.

Figure 3.6 shows modified graph of connections between radicals of genetic code amino acids.

Characteristic of protonic charge of radical of amino acid **His** is [−1 *mod* 4, +3 *mod* 8]. We may suggest that marked amino acid **His** is situated in four-dimensional space-time with signature [−1, +3] and form the surface with Eulerian characteristic $\chi(4) = 0$, because the amino acid **His** (not **Gly**!) is the retract of amino acids (see Figure 3.6). If amino acid **Trp** is unchanged, the protein surface will be in three-dimensional space and will have Eulerian characteristic $\chi(3) = 2$. The difference $\chi(-2) = \chi(4) - \chi(3)$ compensates the signature of generators of Clifford algebra $Cl(4, 2)$.

Compensation of opening of pyrrolidine ring in radical of amino acid **Pro**, with increase of protonic charge of **Trp** amino acid residue by 4 units, results in change of Eulerian characteristic (decrease by two units) of the surface of a protein, composed of genetic code amino acids.

We may directly compensate the signature of Clifford algebra *Cl*(4, 2) through introduction of marked glycine point into characteristic of protonic charge modulo 4. Marked glycine point will have characteristic (–1) and signature of amino acid residues modulo 4 will be equal to [12+ 8–]. This means that amino acid radicals and their pairs form isomorphous Clifford algebras. Two marked points **Gly**{–1 *mod* 4} and **His**{–1 *mod* 4, +3 *mod* 8} determine signature [3+ 2–] of five-dimensional space of protein molecule presentation.

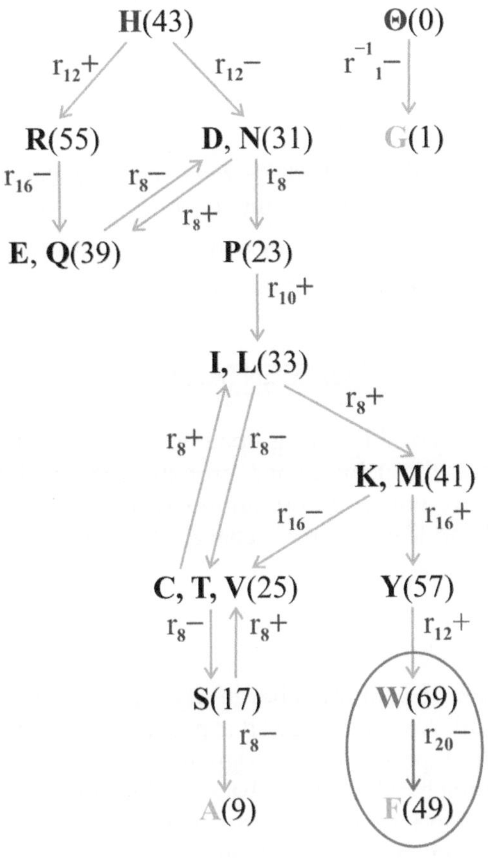

Figure 3.6 Modified graph of links between radicals of genetic code amino acids. Highlighted is the pair of amino acids **W**(69) – **F**(49), which occupied new place under influence of pointed out amino acid **H**(43). Increment of protonic charge is $\chi(-2)$ = Qp[(**F** – **A** + **G**) – **H**].

COMMENTARY 3.2

It is well known that integers, not exceeding a given module N and relatively prime with it, form a group of order φ(N), where φ(N) is Euler numerical function. If N is equal to 2, 4, p^k, $2p^k$, where p is prime odd number, this group is cyclic. Then the cyclic group has φ(φ(N)) generators, that is, proper primitive roots. If a group is noncyclic one, the number N has no proper primitive roots. But it has **improper** primitive roots. To find them, we should do the following:

a) to select for the number φ(N) other module $N_2 = \boldsymbol{p}^k$ at condition φ(N) = $\varphi(N_2)$, that is, to change noncyclic group for cyclic one; or
b) to divide φ(N) into parts, e.g., $\varphi(N) = \varphi(\boldsymbol{p}_1^{k1}) + \varphi(\boldsymbol{p}_2^{k2})$ at condition $\varphi(\varphi(N)) = \varphi(\varphi(\boldsymbol{p}_1^{k1})) + \varphi(\varphi(\boldsymbol{p}_2^{k2}))$, which means the change of noncyclic group for two cyclic ones.

Improper primitive roots generate **code** of noncyclic group.

Let us consider protonic charge values of radicals of genetic code amino acids. Six amino acids, which values Q_p(Amr) are not equal to the power of prime number

$$Q_p(\mathbf{I}) = Q_p(\mathbf{L}) = 33,\ Q_p(\mathbf{E}) = Q_p(\mathbf{Q}) = 39,\ Q_p(\mathbf{R}) = 55,\ Q_p(\mathbf{Y}) = 57,$$

have improper primitive roots:

a) $\varphi(33) = \varphi(\mathbf{5^2}) = 20,\ \varphi(55) = \varphi(\mathbf{41}) = 40,\ \varphi(57) = \varphi(\mathbf{37}) = 36;$
б) $\varphi(39) = \varphi(\mathbf{13}) + \varphi(\mathbf{13}) = 24,\ \varphi(\varphi(39)) = 8,\ \varphi(\varphi(\mathbf{13})) + \varphi(\varphi(\mathbf{13})) = 8.$

In the case that $Q_p(\mathbf{W}) = 73$, Clifford algebra of amino acid radicals, due to belonging of their protonic charge values to the type p^k, is equal to $Cl(14, 6) \cong Cl(11, 9)$. But if $Q_p(\mathbf{W}) = 69$, then b) will satisfied:

$\varphi(69) = 2\varphi(\mathbf{23}) = 44,\ \varphi(\varphi(69)) = 2\varphi(\varphi(\mathbf{23})) = 20$ and Clifford algebra of amino acid radicals, due to belonging of their protonic charge values to the type p^k, is equal to $Cl(13, 7) \cong Cl(9, 11)$.

Selected Bibliography

Stcherbic, V.V., Buchatsky, L.P. The fine-structure constant and structure of protein. Scientific Transactions of Ternopil National Pedagogical University. Ser. Biol. 53: 121–128, 2012 (in Ukrainian).

Biomolecule structure

Moran, L.A., Horton, H.R., Scrimgeour, K.G., Perry, M.D. Principles of biochemistry. New Jersey: Prentice Hall RTR, 2012.

Branden, C., Tooze, J. Introduction to protein structure. 2nd ed. New York: Garland Publishing, Inc., 1999.

Fine-structure constant

Hanneke, D., Fogwell, S. Gabrielse, G. New measurement of the electron magnetic moment and the fine structure constant. Phys. Rev. Lett. 100: 120801, 2008.

Mohr, P.J., Taylor, B.N., Newell, D.B. CODATA recommended values of the fundamental physical constants: 2010. Rev. Mod. Phys. 84: 1527–1605, 2012.

Sommerfeld, A. Zur Quantentheorie der Spektrallinien. Ann. Phys. Lpz. 51: 1–94, 1916.

Dirac, P.A.M. Quantised singularities in electromagnetic field. Proc. R. Soc. (London) A133: 60–72, 1931.

Eddington, A.S. Fundamental Theory. Cambridge: Cambridge University Press, 1946.

Das, A. Coffman, C.V. A class of eigenvalues of the fine-structure constant and internal energy obtained from a class of exact solutions of the combined Klein-Gordon-Maxwell-Einstein Field Equations. J. Math. Phys. 8: 1720–1735, 1967.

Boyer, T.H. Quantum electromagnetic zero-point energy of a conducting spherical shell and the casimir model for a charged particles. Phys. Rev. 174: 1764–1776, 1968.

Kinoshita, T. The fine structure constant. Rep. Prog. Phys. 59: 1459–1492, 1969.

Morris, T.F. Does the fine-structure constant have a dynamic origin. Phys. Lett. 93B: 440–442, 1980.

Chew, G.F. Zero-Entropy bootstrap and the fine-structure constant. Phys. Rev. Lett. 47: 764–767, 1981.

Davies, P.C.W. Constraints on the value of the fine structure constant from gravitational thermodynamics. Int. J. Theor. Phys. 47: 1949–1953, 2008.

Schonfeld, E., Wilde, P. Electron and fine structure constant II. Metrologia 45: 342–355, 2008.

Ross, D.K. An estimation of the fine structure constant using fiber bundles. Int. J. Theor. Phys. 25: 1–6, 1986.

Castro, C. On geometric probability, holography, shilov boundaries and the four physical coupling constants of nature. Progress in Physics 2: 63–69, 2005.

Castro, C. On the coupling constants, geometric probability and complex domain. Progress in Physics 2: 46–53, 2006.

Marek-Crnjac, L. Lie groups hierarhy in connection with the derivation of the inverse electromagnetic fine structure constant from the number of particle-like states 548, 576 and 672. Chaos, Solitons & Fractals 37: 332–336, 2008.

Marek-Crnjac, L. The fundamental coupling constants of physics in connection with the dimension of the special orthogonal and unitary groups. Chaos, Solitons & Fractals 34: 1382–1386, 2007.

Rosen formula

Rosen, G. Group Theoretical basis for the value of the fine structure constant. Phys. Rev. 13D: 830–831, 1976.

Wyler formula

Wyler, A. L'espace symetrique du groupe des equations de Maxwell. C. R. Acad. Sc. Paris 269: 743–745, 1969.

Wyler, A. Les groupes des potentiels de Coulomb et de Yukava. C. R. Acad. Sc. Paris 271: 186–188, 1971.

Cap-structure

Banerjee, A.K. 5′-Terminal cap structure in eukaryotic messenger Ribonucleic acids. Microbiol. Rev. 44: 175–205, 1980.

Clark, D.P. Molecular biology. Understanding the genetic revolution. Elsevier Academic Press, 2005.

Co-evolution theory

Tze-Fei Wong, J. A Co-evolution theory of the origin of the genetic code. Proc. Nat. Acad. Sci. USA 72: 1909–1912, 1975.

Tze-Fei Wong, J. Coevolution of genetic code and amino acid biosynthesis. Trends Biochem. Sci. 6: 33–35, 1981.

Tze-Fei Wong, J. Question 6: Coevolution Theory of the Genetic Code: A Proven Theory. Origins of Life and Evolution of Biospheres 37: 403–408, 2007.

Origin of life

Galimov, E.M. Phenomenon of life: between equilibrium and nonlinearity. Origin and principles of evolution. Moscow: URSS Press, 2001. (In Russian).

Quastler, H. The Emergence of Biological organization. New Haven: Yale University Press, 1964.

Graphene

Nair, R.R., Blake, P., Grigorenko, A.N., Novoselov, K.S., Booth, T.J., Stauber, T., Peres, N.M.R., Geim, A.K. Universal Dynamic Conductivity and Quantized Visible Opacity of Suspended Graphene. arXiv.org/0803.3718, 2008.

Nair, R.R., Blake, P., Grigorenko, A.N., Novoselov, K.S., Booth, T.J., Stauber, T., Peres, N.M.R., Geim, A.K. Fine Structure Constant Defines Visual Transparency of Graphene. Science 320: 1308, 2008.

4

Cancer: Viruses, Attractors, Fractals

Tumors, benign and malignant, may arise in all multicellular organisms. Tumorigenesis is one of the most mysterious problems faced by mankind. Perhaps, no other scientific problem has had so many hypotheses associated with it, as the origin of malignant tumours.

Attempts to understand the mechanism of tumorigenesis have been undertaken from the moment it became clear that tumours are an illness and need to be treated. First theories of tumorigenesis, sometimes speculative and based on some single property of malignant cells, had been appearing in spite of limited knowledge of biological mechanisms of morphogenesis and absence of experimental data on primary biochemical processes. Every new discovery in biology has always been considered as an additional acquirement in the fight against malignant tumours. Physicians have been hoping for the development of a "vaccine against cancer". Cancer, like other diseases, is a natural phenomenon. Tumours have many reasons to arise. Both animals and plants are tormented by tumours. Why is this phenomenon so wide spread? It would seem that during the long period of evolution, a development to prevent the appearance of tumours should have come into existence, however tumours have survived.

Modern scientists who deal with malignant tumours believe that the crucial role in the process of carcinogenesis belongs to genetic disorders. Disorders are multiple, and the more they are, the more malignant are the tumours.

To understand the nature of tumour transformation of normal cells, they have been investigating different factors (chemical, physical, biological) that favour carcinogenesis. Modern theory of carcinogenesis—mutational theory supplemented with the theory of oncogenes and tumour suppressors—

deserves the most attention. Biological factors of oncogenesis are, first of all, oncogenic viruses. Over the years, many oncogenic viruses that can provoke tumours in humans and animals have been discovered. Studying tumour cell transformation under the influence of oncogenic viruses has an advantage compared to chemical or physical cell transformation. Cells transformed by oncogenic viruses have the same phenotype; therefore, with known virus particle structure, the process of neoplastic cell transformation may be accurately traced.

Interaction of oncogenic viruses with cells is well studied. Cell transformation begins with integration of virus genetic material with cell-host genome. Functioning of virus genome is necessary to establish and maintain transformed state of cell. Virus carcinogenesis has a peculiarity: after penetration of virus genome into cell genome, a prolonged latent period could begin, which may continue for many years in natural conditions. So far, we do not know the cause of this phenomenon. Some other factors of carcinogenesis could influence the appearance of oncodisease.

Much attention has been paid to oncogenes and tumour suppressors. It is supposed that all genomes of multicellular organisms contain oncogenes, mutation of which leads to their activation with disturbance of functions regulating cell cycle. Tumour suppressors, being antipodes of oncogenes, prevent emergence of mutant, potentially carcinogenic cells: they cause cell apoptosis, regulate cell cycle and control genome integrity. Virus oncogenes activate cell protooncogenes, but they may also directly control cell transformation (retroviruses) or inactivate tumour suppressors (DNA containing viruses).

Tumours are characterised by uncontrolled cell proliferation. Two types of tumours may be distinguished.

Benign tumours are local and similar in their structure to normal differentiated cells; they grow slowly, may form a capsule of connective tissue around itself, do not intergrow, but move apart adjacent tissues and organs. Sometimes, benign tumours may transform into malignant ones (tumour regeneration).

Malignant tumours are strongly distinguished by their structure from normal cells; they are weakly differentiated, may intergrow in certain tissues and migrate to other parts of the body giving rise to secondary tumours (metastases).

Oncogenesis is of genetic nature: tumour cells have modified, affected genome. The process of tumour regeneration is the result of long accumulation of genetic defects. Genes of proteins, taking part in DNA repair, cell cycle and cell differentiation, are subject to changes.

At that, in spite of the fact that inductors promoting tumor development are well known, a considerable part of tumors arise spontaneously without apparent links with inducing agents.

Tumor transformation lies in the fact that a normal cell acquires the ability to unlimited reproduction and transfers this ability hereditably to daughter cells. Therefore, understanding of carcinogenesis mechanisms is directly connected with the problem of cell division and its regulation.

Carcinogenic transformation of normal cell is accompanied with deep transformations: spectre of functioning genes and synthesizing proteins becomes changed, cytoskeleton changes, surface loses the property of contact inhibition.

Carefully analysing each phase of malignant cell formation, one may try to understand general biological significance of tumours. We believe that investigation of virus oncogenesis is preferable to anything else. Very important is the understanding of the virus particle structure, as we hope that the structure of oncogenic viruses contains additional information about the mechanism of carcinogenic cell transformation. This idea served as a prerequisite for advanced study of polyomaviruses as the most simple of all known oncogenic viruses.

POLYOMAVIRUSES

Polyomaviruses belong to the family of small oncogenic DNA viruses. Polyomaviruses have protein icosahedric capsid of symmetry $T = 7d$. Viral capsids include 72 pentamers and do not contain hexamers. Pentamers of protein capsid are located in the centers of surface lattice. Each pentamer contains five molecules of major protein VP1 and approximately one molecule of each of the minor proteins VP2 и VP3, located nearer to the centre of the virus particle. Arrangement of pentamers is stabilised by calcium ions as well as disulfide bonds between them. VP1 proteins form connected shell of polyomaviruses (Figure 4.1).

The virus genome is a minichromosome that consists of circular DNA, wrapped around nucleosomes of somatic cell. DNA virus contains two types of genes: early genes coding oncoproteins (large and small T-antigens), and late (after replication of virus DNA) genes, which encode proteins of virus capsid. As a result of alternative splicing of primary mRNA transcript, two mRNA are generated that code large and small antigens. Virus DNA replication takes place at obligatory involvement of large T-antigen. Main

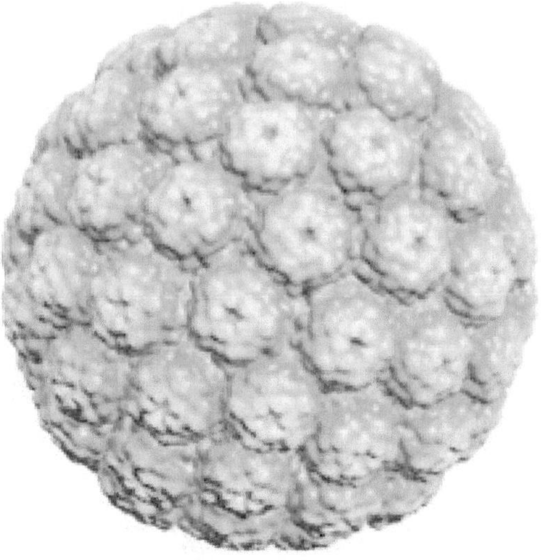

Figure 4.1 Structure of polyomavirus SV40 capsid according to the data of X-ray structure analysis (PDB 1sva).

feature of polyomaviruses is that they produce multifunctional protein, large T-antigen, which is connected with broad spectrum of biological processes: activation of gene transcription and repression, blocking of cell differentiation, stimulation of cell cycle and inhibition of apoptosis.

Genetic research has determined three domains of large T-antigen necessary for transformation. First domain, including 82 amino acids (common also to small antigen), binds with chaperone hsc70, a protein that takes part in assembly and disassembly of protein complexes. Second domain (amino acids 102–115) interacts with tumour suppressors of Rb-family (pRb, p107, p130). pRb protein normally regulates activity of transcriptional factors of E2F-family and cyclin dependent kinases during the beginning of cell cycle S-phase and its successful progress. Large T-antigen dissociates Rb-E2F complex and thereby activates expression of genes of cell growth factors. Loss of pRb function leads to dysfunction of cell differentiation processes with raised possibility of permanently proliferating cell emergence; at accumulation of mutations, this may result in malignant transformation. Third domain (amino acids 351–450, 533–626) interacts with p53 protein, apoptosis inductor, blocking its action at DNA damage.

Through inactivation of the enzyme PP2A, small T-antigen also contributes to cell transformation stimulating growth of noncycling cells.

After incorporation into cell genome, virus DNA replicates together with chromosome DNA. In transformed cell, only polyomavirus T-antigens may express. Proteins of the virus capsid do not express. Total length of large and small T-antigens is approximately equal to half a polyomavirus genome. The second half-genome contains genes of protein capsid.

It seems that oncogenic properties of polyomavirus are somehow bound with structure features of its genome.

Polyomaviruses fall beyond the general pattern of Caspar-Klug theory. The process of polyomavirus capsid formation is rather complex. Arranging of pentamers in accord with symmetry T = 7*d* occurs due to different types of interactions between monomers and their interpenetrating branching. In addition to polyomavirus capsid, VP1 protein may form virus-like structures containing 12 and 24 pentamers.

Skew symmetry T = 3 of polyomavirus protein capsid

Oncogenic properties of polyomaviruses may be connected with specific symmetry of the virus protein capsid. The structure of polyomavirus protein capsid that we propose, is the structure of skew symmetry T = 3, whose parameters are bound with characteristics of genome and capsid proteins.

Symmetries of icosahedron and dodecahedron coincide geometrically, but they are different in respect of filling three-dimensional space. Icosahedron surface may be indefinitely complicated, but it is impossible to form a compact space figure from icosahedrons. On the contrary, various voidless space figures are easily built from dodecahedrons.

Skew symmetry T = 3 is based on expanded dodecahedron (Figure 4.2A), derived by joining 12 dodecahedrons to the initial one. Such capsid

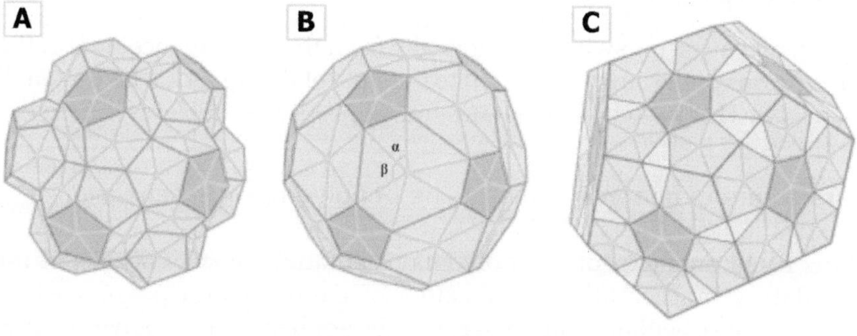

Figure 4.2 Skew symmetry T = 3 of polyomavirus protein capsid.

A) Expanded dodecahedron containing 72 pentamers S5 with the side 1.

B) Convex envelope of expanded dodecahedron. Angles α and β of hexamer S6 are shown.

C) Completely disclosed expanded dodecahedron with free regions S3. Areas of the figures are: S5 = 1.720477, S6 = 4.369171, S3 = 0.293893.

model contains 72 pentamers (360 subunits of VP1 protein), and is not a convex solid. Convex envelope (Figure 4.2B) of expanded dodecahedron has skew symmetry T = 3 and contains 180 protein subunits VP1. In the region of 20 hexamers, this hypothetical envelope contains three molecules each of protein VP2 and VP3.

Therefore, the structure of skew symmetry T = 3 contains protein subunits VP1 two times less then the structure T = *7d*. The numbers of protein subunits VP2 and VP3 in both structures may be assumed coincident.

Principal parameter of the skew symmetry T = 3 is the ratio of central angles in the hexamer (Figure 4.2B):

$$\lambda = \beta/\alpha = 75.5236°/44.4764° = 1.6981.$$

Presentation of polyomavirus protein capsid in the form of three-layered surface structure $\{\mathbf{A} \subset \mathbf{B} \subset \mathrm{C}\}$ depicts more exactly spatial arrangement of capsid proteins that form volumetric (fractal!) grid. Convex envelope of expanded dodecahedron—covering of expanded dodecahedron—plays the role of fractal operator.

STOCHASTIC HENON ATTRACTOR

It is generally assumed that **attractor** is a set of states in which a physical system evolves after a long enough time, regardless of the starting conditions of the system. Chaotic attractor, which we named **stochastic Henon attractor**, will be obtained in a rather unusual way: through identifying roots of Henon map with parameters of a new Henon map. We will show that the pair composed of large and small T-antigens forms stochastic Henon attractor.

Henon attractor switches chaotically from the convergence state of the parts (exons of the large T-antigen) to the state of divergence with the small T-antigen. This switching generates fractal properties of Henon attractor.

Parameters of genome and proteins of polyomavirus capsid

Virus SV40, member of the polyomavirus family, is a basic model for calculating mathematical parameters of genomes and polyomavirus proteins. The calculated parameters may be applied to other polyomavirus to confirm their stability.

We selected polyomaviruses among viruses with not too great deviations for parameters of capsid proteins and genome from the SV40 virus etalon parameters.

The first parameter is the ratio of protein masses:

$$\xi = m(\text{VP3}) / m(\text{VP2}) + 1,$$

where m(VP2) and m(VP3) are masses of proteins VP2 and VP3. Values ξ and λ are assumed to be almost equal (Table 4.1). Protein masses VP2, VP3 were calculated using formulae given in manuals.

The second parameter is connected with length of genes of proteins VP1, VP2 and VP3 and their mutual alignment (Figure 4.3). Gene VP3 is a part of gene VP2. Gene VP1 overlays genes VP2 and VP3 but is read with frame-shift. Main idea when considering genes VP1, VP2 and VP3 is to attempt obtaining such ratio of their lengths, which would be quite close to parameters ξ and λ and should confirm three-layered structure of polyomavirus protein capsid.

The second parameter is calculated by the formula:

$$\nu = \frac{g(\text{Vp2}) + g(\text{Vp3})}{g(\text{Vp1}) - d \times g(\text{Vp3}) / [g(\text{Vp2}) + g(\text{Vp3})]}$$

where g(VP1), g(VP2), g(VP3) are the lengths of genes VP1, VP2, VP3; d is overlay of VP1 gene with VP2 and VP3 genes.

Table 4.1 Parameters of genome and mass of polyomavirus proteins VP2 and VP3.

Polyoma virus	GenBank	Genome b.p.	m(VP2)	m(VP3)	ξ
1. African green monkey	K02562	5270	39324.0	27289.0	1.6940
2. Bovine	D13942	4697	39143.5	26857.4	1.6861
3. BK (Dunlop)	V01108.1	5153	38344.7	26718.3	1.6968
4. JC	J02226	5130	37365.7	25743.3	1.6890
5. Simian 12	NC_012122	5206	38309.9	26841.6	1.7006
6. Simian 40	J02400.1	5243	38524.9	26961.5	1.6999
7. Goose	NC_004800	5256	35108.5	24657.4	1.7023
8. Myotis VM-2008	NC_011310	5081	39949.4	28084.9	1.7030
9. Finch	NC_007923	5278	38346.8	27613.3	1.7201
10. Hpy V6	NC_014406	4926	35967.2	24304.7	1.6757
11. Hpy V7	NC_014407	4952	35563.0	23994.6	1.6747
12. WU	NC_009539	5229	43602.6	30056.1	1.6893

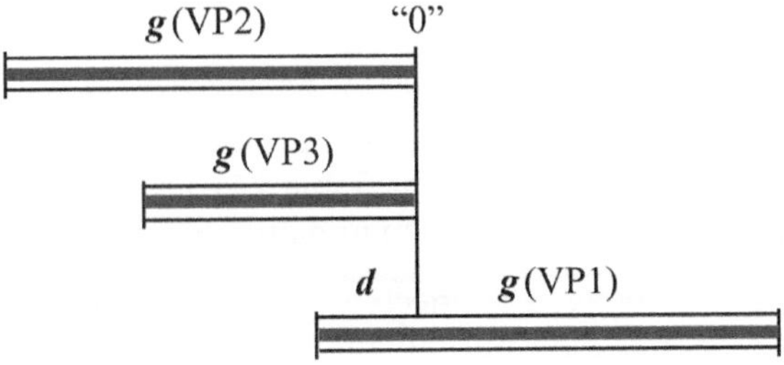

Figure 4.3 Relative position of genes of polyomavirus proteins VP1, VP2 and VP3.

Nonlinear formula for parameter ν was obtained from the necessity to implement most carefully the equality $\nu = \xi$ for SV40 virus. But, rather unexpectedly, this formula led to Henon map.

Parameters of Henon map for polyomavirus proteins

Table 4.2 below shows calculation data for parameters of genes of different polyomavirus proteins.

Table 4.2 Parameters of Henon map of polyomavirus proteins.

Polyoma virus	g(VP1)	g(VP2)	g(VP3)	*d*	ν	Parameters of Henon map		Renormalization	
						μ	*b*	d_2	ν_2
1. African green monkey	1107	1071	714	122	1.6868	5.20426	9.07377	123	1.6875
2. Bovine	1098	1062	699	140	1.6893	4.49039	7.84286	123	1.6785
3. BK (Dunlop)	1089	1056	699	116	1.6830	5.40915	9.38793	122	1.6869
4. JC	1065	1035	678	92	1.6653	6.75517	11.57609	119	1.6829
5. Simian 12	1089	1059	705	116	1.6919	5.39602	9.38793	121	1.6951
6. Simian 40	1095	1059	705	122	1.6860	5.14837	8.97541	122	1.6860
7. Goose	1062	981	654	101	1.6004	6.06891	10.51485	118	1.6112
8. Myotis VM-2008	1074	1062	714	105	1.7213	5.87601	10.22857	119	1.7307
9. Finch	1077	1065	735	104	1.7399	5.88557	10.35577	118	1.7496
10. Hpy V6	1164	1011	648	98	1.4737	7.00019	11.87755	132	1.4913
11. Hpy V7	1143	990	630	71	1.4524	9.60037	16.09859	130	1.4829
12. WU	1110	1243	819	152	1.9690	4.16991	7.302631	124	1.9484

Statistics of polyomavirus protein parameters shows that the best parameter is ξ. Parameter ν has somewhat worse stability.

Suppose that using the formula for calculating parameter ν, we may determine Henon map.

Let us introduce the following designations:

$$g_2 = g(\text{VP2}),\ g_3 = g(\text{VP3}),\ g_1 = g(\text{VP1}).$$

Rewrite the equation for ν as follows:

$$(g_2 + g_3)^2 = \nu g_1(g_2 + g_3) - \nu g_3 d = \nu g_1(g_2 + g_3) - \nu d(g_2 + g_3) + \nu g_2 d.$$

Designate $z = g_2 + g_3$. Then $z^2 = \nu g_1 z - \nu d\, z + \nu g_2 d$ is main equation. Let $z = g_2 x$; $x = (g_2 + g_3)/g_2$. Then $g_2 x^2 = \nu g_1 x - \nu dx + \nu d$. Divide this equation by νd:

$$-(g_2/\nu d)\, x^2 - x + (g_1/d)\, x + 1 = 0 \text{ – equation of Henon map.}$$

Let us write Henon map in standard form:

$$x_{n+1} = 1 - \mu x_n^2 + y_n$$
$$y_{n+1} = b x_n$$

where parameters of Henon map are: $\mu = g_2/\nu d$, $b = g_1/d$. Value x_n is the main variable and y_n is an auxiliary one since $y_n = bx_{n-1}$. Fixed points of this map are $x_1 = 1.66572$, $x_2 = -0.116608$. Point x_1 determines virus protein shell since $x_1 \approx \nu_{average} = 1.6716$. Both fixed points are unstable. Note that value x makes sense over a period when expression of VP2 and VP3 genes takes place.

Henon map parameters μ and b for polyomaviruses differ somewhat. We may introduce parameters $\mu_0 = 5.148367$ and $b_0 = 8.97541$ of universal Henon map for all polyomaviruses, which coincide with parameters μ and b for SV40 virus. This may be done if one takes into account that the **cause of divergence of Henon map parameters from their universal values is correlation between gene length and its "numerical value"**. Varying *only one parameter d* we may achieve *coincidence* of parameters μ and b among selected 12 polyomaviruses.

New, renormalized value ν_2 is calculated from:

$$\nu_2 = [g_2 + g_3 + \{g_2 \times g_3/(g_2 + g_3)\}/\mu_0]/g_1.$$

Then new, renormalized *integer* value $d_2 = (g_2/\mu_0)/\nu_2$ is determined and, over again, parameters μ and b are determined, which almost coincide with μ_0 and b_0.

Henon map parameters for antigens

Figure 4.4 shows the structure of large and small T-antigens of polyomaviruses.

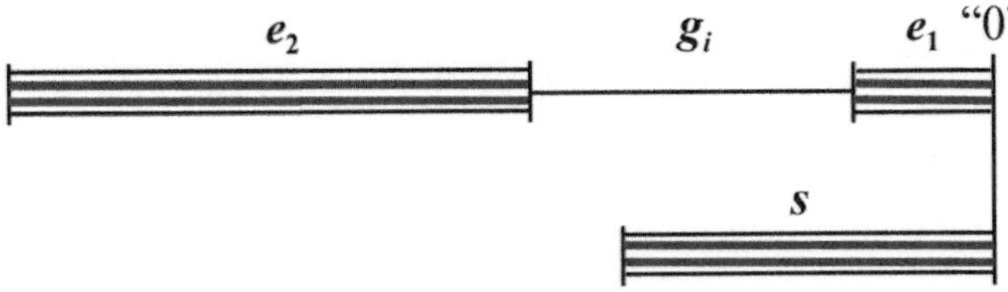

Figure 4.4 Structure of large and small T-antigens of polyomaviruses. Large T-antigen is a sum of two exons e_1 and e_2. Left end of small antigen s is inside intron g_i of large T-antigen. Large and small T-antigens are read from the right on the left.

Suppose that parameters of Henon map μ and b for polyomavirus protein capsid almost coincide with parameters of Henon map μ_e и b_e, which are determined by the structure of large and small T-antigens. However, in the structure of large and small T-antigens, DNA segment equivalent to the gene g_3 is absent. Therefore, in this case the formula for parameter ν changes and will be:

$$\nu_e = \frac{N_0}{g_1' - e_1 + [g_2' \times e_1] / N_0}$$

where $g_1' = e_2 + s$;

$g_2' = e_2 + g_i$;

N_0 is the volume of genetic information necessary for coding nucleosome.

Here we are faced with large uncertainty: while genes of histones H2A, H2B, H3 and H4 are stable lengthwise, gene of histone H1 has no fixed value. Nominal value is:

$$N_0 = 2 \times g(\text{H2A}) + 2 \times g(\text{H2B}) + 2 \times g(\text{H3}) + 2 \times g(\text{H4}) + g(\text{H1}) =$$
$$= 2 \times 387 + 2 \times 375 + 2 \times 405 + 2 \times 306 + 582 = 3528 \text{ n.}$$

Relative to the structure of genes VP1, VP2 and VP3, the structure of large and small T-antigens differs for the reason that shift e_2 relative to "0" is left, while shift g(VP1) relative to "0" is right.

Parameters of Henon map for antigens: $\mu_e = g_2'/[\nu_e \times e_1]$, $b_e = g_1'/e_1$. The variable x_e is determined as N_0/g_2' and makes sense, when expression of initial transcript of large and small T-antigens in **the structure of cell genome** takes place. Parameters of large and small T-antigens of polyomaviruses are delivered in Table 4.3.

Table 4.3 Parameters of large and small T-antigens of polyomaviruses.

Polyoma virus	g(LTag)	s g(sTag)	g_1'	g_2'	g_i	g(LTag) / g(sTag)	m(LTag) / m(sTag)	φ(LT)	φ(A)
1. African green monkey	2091	567	2422	2211	356	3.6878	3.6078	0.7273	1.5590
2. Bovine	1758	372	2026	1902		4.7258	5.0440	0.6528	1.5128
3. BK (Dunlop)	2085	516	2359	2188	248 + 72	4.0407	3.9357	0.7344	1.6051
4. JC	2064	516	2338	2167	345	4.0000	3.9190	0.7336	1.5967
5. Simian 12	2097	516	2371	2204	345	4.0639	3.9581	0.7329	1.6026
6. Simian 40	2124	522	2401	2226	349	4.0689	3.9916	0.7356	1.6147
7. Goose	1908	480	2092	1805	347	3.9750	3.8926	0.8286	1.9354
8. Myotis VM-2008	2010	486	2260	2048	193	4.1358	4.1733	0.7656	1.7389
9. Finch	1836	498	2032	1739	274	3.6867	3.5658	0.8199	1.8464
10. Hpy V6	2007	570	2335	2156	391	3.5210	3.3839	0.7110	1.4675
11. Hpy V7	2013	579	2347	2174	406	3.3575	3.3103	0.7060	1.4440
12. WU	1944	582	2275	2093	400	3.4587	3.1738	0.7078	1.4246
				Mean value		3.8936	3.8297	0.7380	1.6123

Renormalization of values μ_e and b_e is performed by varying two parameters N_0 and e_1. Let $\gamma_0 = b_0/\mu_0 = 1.74335$. Then, knowing values μ_0 and b_0, we may calculate $\nu_{0e} = [\gamma_0 \times g_2']/g_1'$ and $e_{10} = g_2'/[\mu_0 \times \nu_{0e}]$.

Then there may be determined renormalized value N_e from the equation:

$$N_e^2 - \nu_{0e} \times (g_1' - e_{10}) \times N_e - \nu_{0e} \times e_{10} \times g_2' = 0.$$

The value x_e is redefined as N_e/g_2'. Positive root $x_{e1} = x_1$ supposes coincidence of reading direction for large T-antigen and genes of nucleosome histones.

Structure parameters of large and small T-antigen should contain information about skew symmetry T = 3 of polyomavirus protein capsid, as only T-antigens of the virus can express in transformed cell.

Let us introduce parameter σ nearly equal to ξ:

$$\sigma = [g(\text{Tag}) + 2 \times g_i] / \text{N}_0,$$

where $g(\text{Tag}) = 2 \times [g(\text{LTag}) + g(\text{sTag})]$;

$g(\text{LTag})$, $g(\text{sTag})$—lengths of large and small T-antigens.

For SV40 virus, we will have:

$$g(\text{Tag}) = 2 \times [2124 + 522] = 5292 \text{ b.p.};$$
$$\sigma = \{5292 + 2 \times 347\}/3528 = 1.6939.$$

We regard intron g_i as **supplement** (not a part!) of the virus genetic information. Equality of parameters ξ = σ requires renormalization of the values:

$$\text{N}_0 \to \text{N}_e \text{ and } g(\text{Tag}) \to g_e(\text{Tag}) = \xi \times \text{N}_e - 2 \times g_i.$$

Table 4.4 contains data and results of calculation for parameters of Henon map for T-antigens of polyomaviruses.

Table 4.4 Parameters of Henon map for large and small T-antigens of polyomaviruses.

Polyoma virus	e_1	ν_e	g(Tag)	Parameters of Henon map		Renormalization			
				μ_e	b_e	e_{10}	ν_{0e}	N_e	g_e(Tag)
1. African green monkey	236	1.5116	5316	6.19770	10.26271	270	1.5916	3683	5527
2. Bovine	104	1.7836	4260	10.25392	19.48077	226	1.6366	3168	4701
3. BK (Dunlop)	242	1.5562	5202	5.80993	9.74793	263	1.6167	3644	5493
4. JC	242	1.5717	5160	5.69721	9.66116	260	1.6159	3610	5407
5. Simian 12	242	1.5472	5226	5.88623	9.79752	264	1.6204	3671	5544
6. Simian 40	245	1.5269	5292	5.95048	9.80000	268	1.6165	3708	5609
7. Goose	296	1.8117	4776	3.36605	7.06757	233	1.5044	3007	4732
8. Myotis VM-2008	236	1.6326	4992	5.31550	9.57627	252	1.5797	3411	5260
9. Finch	302	1.8777	4668	3.06661	6.72848	226	1.4920	2897	4573
10. Hpy V6	242	1.5744	5154	5.65881	9.64876	260	1.6095	3591	5235
11. Hpy V7	245	1.5659	5184	5.66658	9.57959	261	1.6151	3622	5253
12. WU	251	1.6236	5052	5.13580	9.06375	253	1.6041	3487	5090

Parameters ν_2 and ν_{0e} are different; they may be done coincident, *if it is possible* to renormalize values $g_2 \to [\nu_{0e} \times g_2] / \nu_2$. This means that parameters of protein genes and parameters of large and small T-antigens may be entirely equivalent. In this case, oncogenic cell transformation is possible but has low probability because of large latent period.

It may be accepted that exponential divergences of Henon map for capsid proteins of polyomaviruses and T-antigens alternately overtake each other. Then we may affirm that there is some set of their mutual attraction—attractor. Exactly stochastic Henon attractor is this attractor.

Henon attractor

Suppose that cell transition from normal to transformed state is bifurcation with Henon map parameters, which are equal to fixed points of universal Henon map:

$$\mu = x_1 = 1.66572, b = x_2 = -0.116608.$$

These parameters determine Henon attractor (Figure 4.5).

Henon map has fixed points $x_1 = 0.5090319$, $x_2 = -1.179377$, each of which is partially steady.

On Figure 4.5, successive points of Henon attractor bind region $L1_3$ with two regions L2 and $L1_1$. Comparison of regions L1 and L2 with similar ones in classic Henon map (Figure 4.5) shows that directions of orderliness of L1 and L2 regions (as sets) coincide.

Henon attractor is a core set, generator of *mutually limited* recession of trajectories of molecules that were primarily disposed in a local region of cellular space.

Let us compare Figures 4.4 and 4.5. It would seem that we may easily build correspondence of Henon attractor sets to the structural elements of large and small T-antigens, that is, $L1_1$, $L1_2$, $L1_3$, L2 ~ e_1, g_i, e_2, s. No, that is not so! To find adequate correspondence, we will calculate for 12 polyomaviruses mean value of the ratio of gene lengths [g(LTag) = e]/[g(sTag) = s] and mean value of mass ratio for proteins of large and small T-antigens [m(LTag)]/[m(sTag)] (Table 4.3).

Assume that $\Lambda 1_1$, $\Lambda 1_2$, $\Lambda 1_3$, $\Lambda 2$ are the lengths of corresponding L-sets as lengthwise curved lines. In chosen coordinate grid, we will have:

$$\Lambda 1_1 = 461.3351, \Lambda 1_2 = 258.2421, \Lambda 1_3 = 465.2161, \Lambda 2 = 306.5110.$$

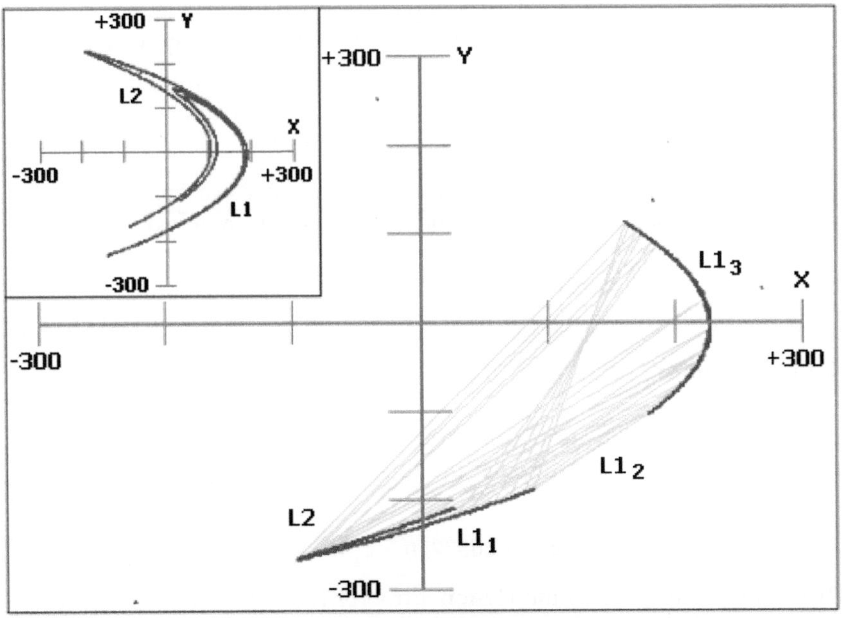

Figure 4.5 Henon attractor at μ = 1.66572, b = – 0.116608 after 10^6 iterations. Values $X = 250x_n$, $Y = 2500y_n$. There are shown selective right lines that bind regions of L1 and L2 switching. Classic Henon map ("horseshoe") with parameters μ = 1.4, b = 0.3, $X = 150x_n$, $Y = 600y_n$ is shown in the left upper corner.

Ratio $\theta(A) = [\Lambda1_1 + \Lambda1_2 + \Lambda1_3]/\Lambda2$ is equal to 3.8654 and almost coincides with mean values of ratios [*g*(LTag)]/[*g*(sTag)] and [*m*(LTag)]/[*m*(sTag)] in Table 4.3. Hence, it is clear that correct correspondence is $(L1_1 + L1_2 + L1_3) \sim (e_1 + e_2)$, L2 ~ *s*, which is rather unexpected.

Such correspondence of the attractor ranges to T-antigens assumes self-action of Henon attractor: **alternative splicing of mRNA primary transcript of large and small T-antigens gives rise to Henon attractor.**

Thus, we demonstrated that the pair – large and small T-antigens—is Henon attractor. It is well known that an attractor can appear in dissipative physical systems.

Figure 4.6 shows switching of Henon attractor ranges. Certainly, one may identify the attractor ranges $L1_1$ ~S106…T124 and $L1_3$ ~ S639…T701 with amino acid phosphorylation regions of large T-antigen of SV40 virus on the basis that ratios *g*(LTag)/*g*(sTag) and *m*(LTag)/*m*(sTag) are very close. But, because of unusual properties of large T-antigen, namely, existence

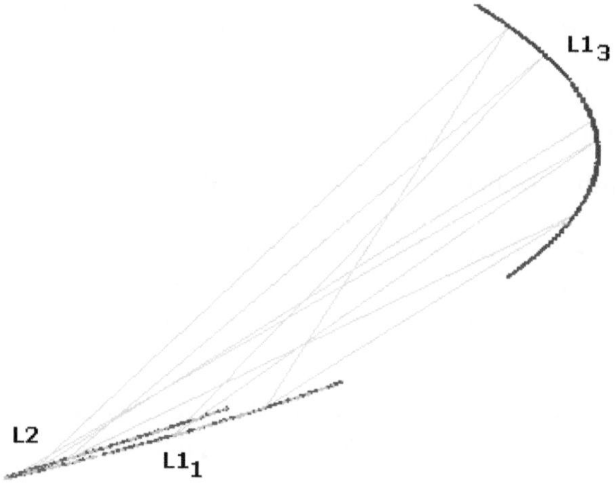

Figure 4.6 Mutual mapping of Henon attractor ranges.
It is shown in colour switching points of the attractor sets modulo 4: 1 *mod* 4 = *green;* 3 *mod* 4 = *red;* 0, 2 *mod* 4 = *blue.*

of two p53-regions of amino acids 351–450, 533–626, we fail to obtain unambiguous interpretation. Possibly, it is the manifestation of stochasticity of large T-antigen.

Projections $L2 \rightarrow L1_3$ and $L1_1 \rightarrow L1_3$ each have one-way direction. There are two directions from the $L1_3$ set: $L1_3 \rightarrow L2$, $L1_3 \rightarrow L1_1$ and they are stochastic. Such a situation is typical for mRNA splicing. Intron rejection is dissipative biological process.

Figure 4.7 is an addition to the Figure 4.6. It shows time sequence of switching between ranges L2 and $L1_1$.

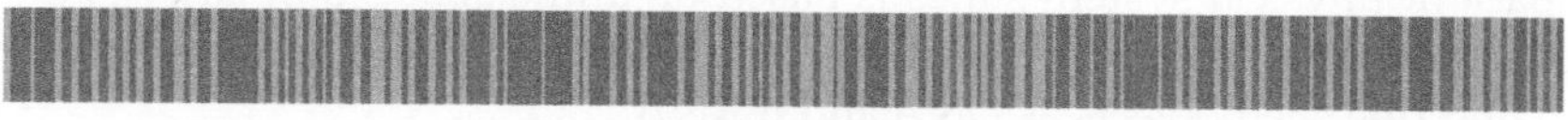

Figure 4.7 Stochastic sequence of switching between ranges of Henon attractor.
$L1_1$ = *black;* L2 = *grey;* 250 numerical values.

The simplest interpretation of mRNA splicing may be easily obtained in affine space—this is the result of two perspective but not mutually-reverse transformations. However, in such an approach there is no stochasticity. Besides, occurrence of cap-structure and poly-A in mature mRNA requires expansion of affine space. Fractal space of states has larger possibilities for description of mRNA splicing.

Henon attractor possesses a very important property: ranges L2 and $L1_1$ will not be directly connected between themselves, if starting point of attractor construction is rejected. Stochastic character of switching between L2 and $L1_1$ indicates that the starting point is quickly forgotten and has no effect on the *bifurcation* properties of the attractor.

Henon attractor behaviour is unpredictable, stochastic, even when it forms for the first time on basis of T-antigen expression of virus circular DNA, as the link between L2 and $L1_1$ ranges is very weak. This leads to the absence of large T-antigen (as initiator) at replication of virus DNA, but gives the last the chance to penetrate into cell genome.

FRACTAL PROPERTIES OF T-ANTIGENS AND PROTEIN CAPSID OF POLYOMAVIRUSES

Stochastic attractor is a fractal. At arbitrary increase, Henon attractor maintains its proportions. Attractor properties of T-antigens are not so obvious.

We will remember how fractals are defined. Let N be the number of some parts of a selected structure, and let $r < 1$ be the coefficient of similarity of a part to all the structure. Then, $Nr^d = 1$. If d is an integer, then almost always it will be a topological dimension of the structure. But if d is a non-integer, then the structure will be fractal. Mathematical fractal is a geometric figure, some part of which is repeated again and again that constitutes one of fractal properties—self-similarity. Non integrity, fractionality of fractal distinguishes it from objects of classical geometry. Fractals are objects, which cannot be simplified. Tangled protein configurations in virus capsid are closer to fractals then to simple icosahedric surfaces.

Hilbert space of functions is based on Euclidean vector space. Fractal space of physical system states is based on affine space. Similarity is the main property of affine space. But even this property should be extended to describe such objects as DNA and mRNA.

Similarity of two structures λ and η, in terms of their fractal properties, may be characterized, separating their structural unities, by the formula:

$$\theta = \sum \lambda_i = \sum (\eta_j)^{\varphi},$$

where θ is the coefficient of structure similarity;

λ_i are topological structural coefficients;
η_j—fractal structural coefficients;
φ—index of fractality (φ-factor) of the structure η.

φ-factor is expansion of fractional index *d*; it may possess even negative values and may be absolute or relative. Fractal structural properties reveal themselves most sharply at introducing of relative (unobvious) φ-factor.

If θ = 1, then φ is dimension of attractor similarity. φ-factor of sparse structures is always less then 1; layered and compacted structures have φ-factor more then 1.

Relative to primary transcript of mRNA, its splicing has $\varphi < 1$ and is defined by weighting coefficients ($e_i < 1$) of exon inclusion into mature mRNA: $\sum_i (e_i)^{\varphi} = 1$. The question how much exons are similar to the primary transcript is open-ended. Nevertheless, mature mRNA may be considered to be an attractor of mRNA primary transcript. Besides, splicing happens once without further recurrence. Therefore, we may assume that *self-similarity of mRNA fractal occurs in the internal fractal space of mRNA states.*

φ-factor of large and small T-antigens

Large and small T-antigens are oncogenes because for both of them $\varphi > 1$. To prove this, we will calculate φ(L1) for L1 range and φ(H) for Henon attractor. To calculate φ(L1), we have the equation:

$$(\Lambda 1_1)^{\varphi\,(L1)} + (\Lambda 1_3)^{\varphi\,(L1)} = (\Lambda 1_1 + \Lambda 1_2 + \Lambda 1_3)^{\varphi\,(L1)}.$$

Calculation gives φ(L1) = 0.7381. To calculate φ(H), we will assume that mRNA of small T-antigen is mRNA attractor of large T-antigen and, accordingly, L2 range is attractor of L1 range. Then, φ(H) is the relative φ-factor, which satisfies the equation:

$$[(\Lambda 1_1)^{\varphi\,(H)} + (\Lambda 1_3)^{\varphi\,(H)}]/(\Lambda 2)^{\varphi\,(H)} = \theta(A) = 3.8654.$$

From here we find φ(H) = 1.5952 > 1. Further, considering the sum of exons e_1 and e_2 as attractor of primary mRNA transcript, we calculate φ(LT) for mRNA of large T-antigen (Table 4.3). In this case, index φ(LT) is the dimension of similarity of large T-antigen mRNA to primary mRNA transcript:

$$(e_1)^{\varphi\,(LT)} + (e_2)^{\varphi\,(LT)} = (e_1 + e_2 + g_i)^{\varphi\,(LT)}.$$

Mean value $\varphi(LT)_{mean} = 0.7380$ coincides with φ(L1).

Formula for calculating φ(A) of large T-antigen mRNA relatively small T-antigen mRNA, chosen by results of experimental data (Table 4.3), is as follows:

$$\frac{e_1 + e_2}{s} = \frac{(s + e_1 + 2g_i)^{\varphi(A)}}{s^{\varphi(A)} + (e_1)^{\varphi(A)}}$$

Mean value $\varphi(A)_{mean}$ = 1.6123 almost coincides with φ(H). These results confirm once more that large and small T-antigens generate Henon attractor.

φ-factor of polyomavirus protein capsid

Let us calculate theoretical φ-factor of polyomavirus protein capsid based on model structures shown on Figure 4.2. Convex envelope of extended dodecahedron contains 12 pentamers S_5 and 20 hexamers S_6. At complete opening of extended dodecahedron, 72 pentamers S_5 and 60 triangles S_3 are formed. Extended dodecahedron is the attractor of completely opened extended dodecahedron. Convex envelope of skew symmetry T = 3 and completely opened extended dodecahedron form two layers over extended dodecahedron. Therefore, theoretical φ(P) of polyomavirus protein capsid must be more than 1.

Ratio $\theta(P) = (72\,S_5 + 60\,S_3)/(12\,S_5 + 20\,S_6) = 1.309905$ is the principal one. Extended dodecahedron is connected structure containing 72 pentamers. Three-layered shell (Figure 4.2) has two relative φ-factors, which may be found as solutions of the equation:

$$\theta(P) = \frac{(72S_5)^{\varphi(P1)}}{(12S_5 + 20S_6)^{\varphi(P1)}} = \frac{(72S_5 + 60S_3)^{\varphi(P2)}}{(72S_5)^{\varphi(P2)}}$$

Accordingly, we have φ(P1) = 1.97239, φ(P2) = 2.02839. It may be supposed that on the average both φ-factors are equal to 2, as in Brownian particle trajectory on plane.

Protein shell of polyomavirus has two layers: VP1 protein subunits are arranged over subunits of proteins VP2 and VP3. Therefore, one may suppose that φ-factor of polyomavirus protein shell is equal to 2.

MALIGNANT TUMOR IS A GIGANTIC COMPACTION OF THE FRACTAL STATE SPACE

Fractal dimension allows evaluating degree of complexity and entanglement of biological structures. Presumably, polyomavirus protein capsid is equivalent to Brownian chaos. For normal fractals, φ-factor of any structure slightly differs from its topological dimension. But, in the space of fractal states, at defined number of identical biological structures, condensation of states is possible. Degree of identity of the structures may not be too large: of importance is only proximity of their φ-factors.

There may be superposition of fractal states, when similarity factor becomes normalized by the number of structures—summation of fractal states. This operation is applied for strongly bound structures. Multiplication of fractal states is applied for weakly bound structures.

Compaction (condensation) of fractal states has a natural basis—condensation of chromosomes. During mitosis, interphase chromosomes are subjected to strong condensation; their linear size may decrease by $k_H = 100$ times with formation of metaphase chromosomes. Value k_H is the main relative parameter of reversible chromosome condensation (decondensation). Suppose that **condensation of fractal states is irreversible at $k_H > 100$.**

Significant circumstance is that interphase chromosomes remain in nucleus; that is, they are surrounded by membrane, while metaphase chromosomes have no envelope. Let us assume that φ-factor for metaphase chromosome takes positive values, while φ-factor for interphase chromosome is negative. Then k_H is modulus of ratio of metaphase chromosome φ-factor to φ-factor of interphase chromosome.

Definition. Benign tumours in (pseudo) envelope have negative φ-factor, without envelope – φ-factor will be positive. Malignant tumours have no envelope. They have positive φ-factor, while metastases have negative φ-factor.

Most biological structures are built on basis of polymer chains. If such structures have φ-factor more then 1, they are potential generators of compaction of fractal states. Large and small T-antigens are generators of oncogenic transformation of normal cells.

Function of compaction of fractal states

Exponential growth function is often used at determining fractal index. But equivalence function increases even more quickly. Exactly, **factorial is a function of condensation of fractal states.**

It is known that $n! \approx e^{-n}n^n(2\pi n)^{1/2}$. At normalization on equivalence function, superexponential nLOG(n) increase of fractal state density appears. This leads to development of a tumour. Type of tumour depends on the value of its φ-factor.

Figure 4.8 shows two graphs of φ-factor increase depending on the number of equivalent cells. Each cell is modelled by two identical parameters equal to 1/3 on condition that

$$(1/3)^{\varphi} + (1/3)^{\varphi} = 1 \text{ (Cantor dust).}$$

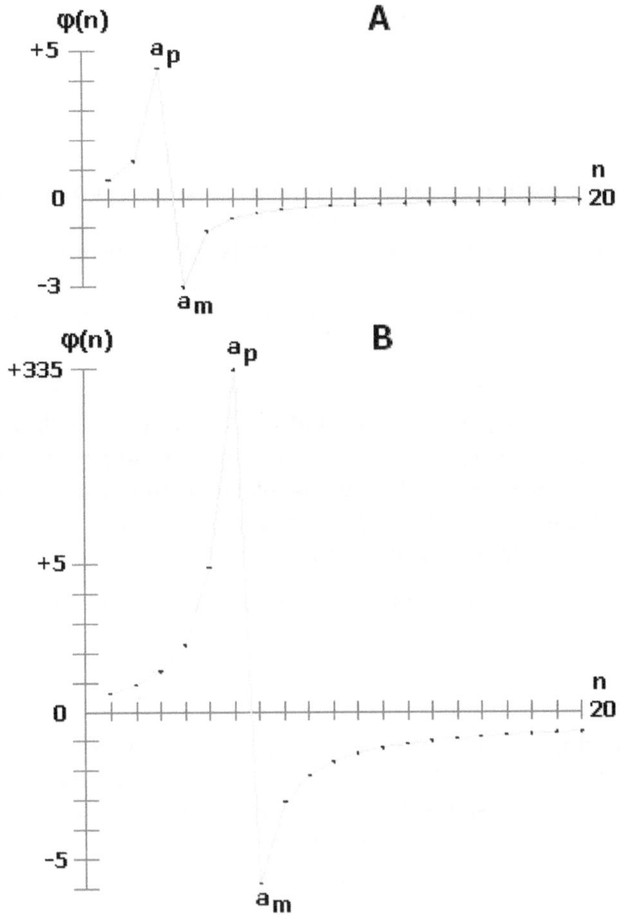

Figure 4.8 Increase of φ-factor depending on the number of equivalent cells. (**A**) Sum of fractal states (benign tumour). (**B**) Product of fractal states (malignant tumour). Each tumour is characterised by the ratio $k_H = -a_p/a_m$ of positive to negative maximum of φ-factor.

The sum of fractal states of n cells is normalized on $n/n!$. Product of fractal states of n cells is normalized on $1/n!$

For a benign tumour, the process of its growth (inflation) is accompanied by insignificant increase of φ-factor; then the tumour passes into the phase of capsulation with negative φ-factor.

It's more complicated with a malignant tumour. Owing to high value k_H, cell malignancy may be reversible or irreversible. Numerical value $k_H = 100$ may be considered as very conditional. Resorption of malignant tumour is a virtual metastasis, as malignant tumour cannot have a real envelope. On the contrary, the real metastases have only briefly negative φ-factor, which varies rapidly over a large positive φ-factor. Real φ-factor for malignant tumour reaches gigantic values, which indicates incredible entanglement of the tumour. No wonder that effective methods for treating oncologic diseases in such situations are absent.

The ratio $k_H = -a_p/a_m$ may take broad spectrum of values for both benign and malignant tumours. Small k_H values are not typical for malignant tumour. **Equivalence function will compact arbitrary number of fractal states,** no matter how parameters are used to describe fractal space of real cell states.

Resonances of equivalence for benign tumour

Assume that each cell is modeled by two integer parameters h on condition that $(1/h)^\varphi + (1/h)^\varphi = 1$. Value $2/h$ shows a part of cell fractal states responsible for tumour growth. Even such a simple model gives a very large variety of cases for different tumours. For benign tumour, integer value h resonates with factorial function. To eliminate resonance denominator (equivalence denominator), fractalization of the h value is used, that is, its slight deviation from integer value is taken into consideration: $\mathrm{LOG}\{h/(n-1)!\} = 0.001$. **Resonances of equivalence for benign tumour may transfer it into a malignant one**. Malignant tumour has no resonances of equivalence.

Figure 4.9 shows dependences of maxima a_p and a_m for different h values.

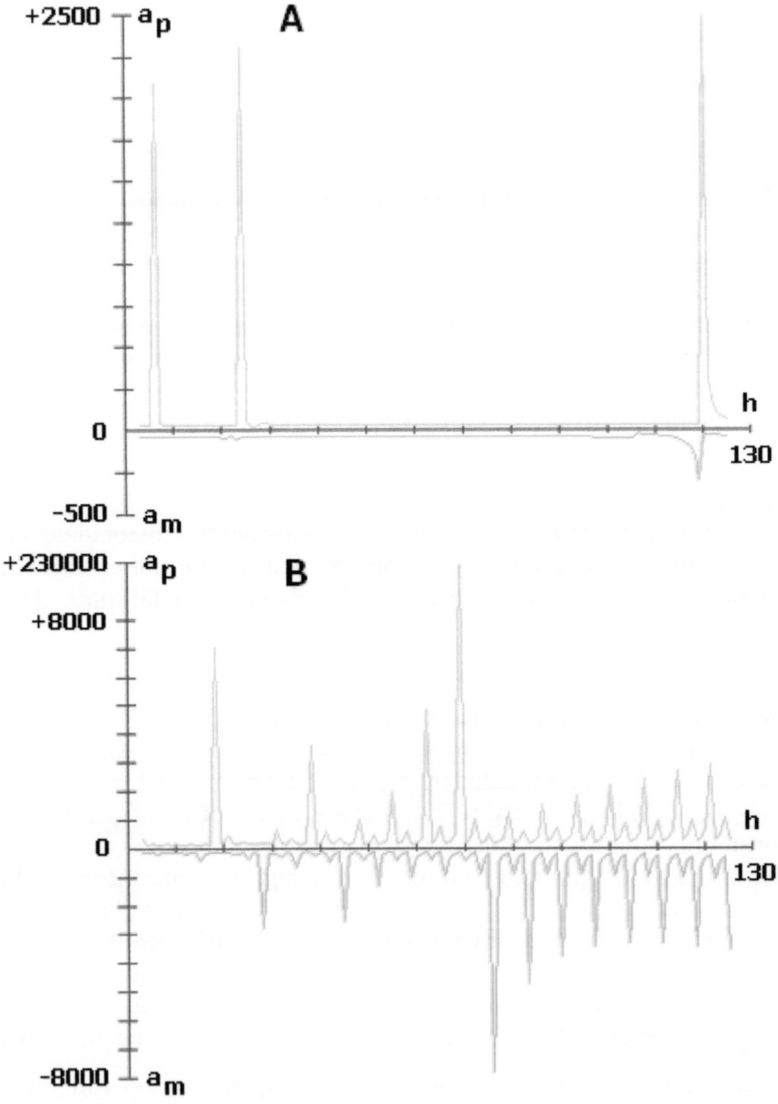

Figure 4.9 Maxima a_p и a_m for different h values.

A) Benign tumour. Resonances of equivalence are observed at $h = 6, 24, 120$. Formula of state condensation: $2n = [h/(n-1)!]^{\varphi}$.

B) Malignant tumour. Typical values: $a_p(69) = 228509$, $k_H(69) = 1758$, $a_m(76) = -7862$, $k_H(76) = 1/54$. Formula of state condensation: $2^n = [h^n/n!]^{\varphi}$.

Condensation of fractal states

There are no strictly identical cells. Molecules of substances in each cell are in permanent motion; multiple chemical reactions occur. Quantitative equalization of the cell molecular content takes place after mitosis. Mitosis ensures uniform distribution of hereditary material between daughter cells; it itself is a rather complicated process of structural transformations and displacement of cell components. Nevertheless, one may affirm that φ-factor of mitotic cell decreases. Formation of cleavage spindle is a unique biochemical process, which requires highly ordered movement of protein molecules. However, at ordered molecular movement, φ-factor is equal to one; that is, it is too small. It is clear that every ordered molecular movement decreases φ-factor. We may affirm that division of a cell is its reaction to the growth of φ-factor (chaotic molecular movement always exists).

Increase of φ-factor (condensation of fractal states) occurs due to disintegration of nuclear envelope and condensation of chromosomes. Cell tends to preserve its φ-factor, its complexity and entanglement. Tumour cells differ from the normal ones in many parameters and are characterized by different structural disorders. Tumour cells divide continuously and their φ-factor is greater then in normal cells.

It is obvious that condensation of fractal states has its limit. But, it is not known so far what it is like. Increase of φ-factor is a kind of cell inflation, which follows with cell discharge, that is, cell division. Simple cell division is characterized by φ-factor equal to 1. Subsequent division of cells results in increase of their φ-factor (cell differentiation). **Tumour may be defined as stochastically differentiated new formation.**

CLIFFORD ALGEBRA OF TUMOUR GROWTH

For benign tumour, φ-factor of condensation of fractal states is normalized on double number of cells $2n$. Far from resonance, values a_{p} and a_{m} satisfy the relation $a_{\mathrm{p}} - a_{\mathrm{m}} = 2n^*$, were n^* is the number of cells (marked point), at which sign of φ-factor changes.

It is well known that Grassmann algebra is equivalent to Clifford algebra with doubled number of generators. Assume that in marked point, vector space of dimension 2^{n^*} with Grassmann algebra is generated. Grassmann algebra has its natural basis in RNA (Chapter 2). One may assume that RNA molecule is responsible for consolidation of fractal states of benign tumour.

At resonance, the number of generators of Grassmann algebra gets up to 1000 and more, but is inaccessible for accurate estimate. In this case, the number of generators of Grassmann algebra significantly exceeds n^*.

In malignant tumour, the arrangement of vector space in marked point n^* is more complicated; in this case, φ-factor of consolidation of fractal states is normalized by growth factor 2^n. Marked point (pole) is determined from the equation $h^{n^*} = n^*!$ Whole family of Clifford algebras is generated in the marked point. Main Clifford algebra is equal to:

$$Cl(a_p^+, -a_m^-), \text{ where } a_p^+ = \mathrm{INT}(a_p) + 1, a_m^- = \mathrm{INT}(a_m).$$

In the pole, φ-factor discontinues. Difference $(a_p^+ - a_m^-)$ differs greatly from INT(n^*).

The following relation is performed with high accuracy:

$$n^* \approx div = \mathrm{INT}(n^*) - a_m^- / (a_p^+ - a_m^-),$$

which testifies that Clifford algebra $Cl(a_p^+, -a_m^-)$ is a *covering* of the growth factor in the pole n^*.

Besides main Clifford algebra, Clifford subalgebras of smaller dimensions $Cl(a_p^+(-k), -a_m^-(+k))$ are generated in the vicinity of the pole; they are defined symmetrically relative to the pole in points INT(n^*) – k, INT(n^*) + k + 1. These subalgebras are present independently and are not a part of main Clifford algebra. So, *the whole tangle of Clifford algebras is generated.*

Clifford algebra has natural basis in DNA (Chapter 1). We may consider that DNA molecule is responsible for consolidation of fractal states of malignant tumour.

Table 4.5 gives values a_p^+, a_m^-, n^*, div for different values of h parameter. Five first Clifford subalgebras at some values of h parameter are given in Table 4.6.

Table 4.5 Values a_p^+, a_m^-, n^*, *div* for malignant tumour.

h	a_p^+	a_m^-	n^*	div
3	335	–6	6.0161	6.0176
4	12	–20	8.6152	8.625
5	38	–13	11.2389	11.2549
6	13	–92	13.8787	13.87619
7	24	–28	16.5298	16.53846
8	78	–19	19.1892	19.19588
9	20	–116	21.8552	21.85294
10	35	–40	24.5264	24.53333

Table 4.5 contd....

Table 4.5 contd.

h	a_p^+	a_m^-	n^*	div
11	101	–26	27.2018	27.20473
12	25	–188	29.8808	29.88263
13	43	–56	32.5627	32.56565
14	106	–35	35.2472	35.24823
15	30	–423	37.9339	37.93377
16	48	–80	40.6226	40.625
17	102	–47	43.3129	43.31544
18	7046	–35	46.0048	46.00494
19	51	–118	48.6981	48.69822
20	95	–62	51.3926	51.39491
21	446	–44	54.0882	54.08979
22	53	–192	56.7848	56.78367
23	89	–84	59.4823	59.48555
24	249	–56	62.1807	62.18361
25	53	–390	64.8799	64.88036
26	84	–117	67.5797	67.58209
27	181	–71	70.2803	70.28175
28	54	–2824	72.9815	72.98124
29	80	–172	75.6832	75.68254
30	146	–92	78.3855	78.38655
31	658	–65	81.0883	81.0899
32	76	–289	83.7916	83.79178
33	125	–123	86.4953	86.49596
34	320	–80	89.1995	89.2
35	73	–684	91.904	91.90356
36	111	–173	94.6089	94.60915
37	221	–102	97.3142	97.31579
38	3603	–74	100.0198	100.0201
39	101	–268	102.7258	102.7263
40	174	–133	105.432	105.4332

Table 4.5 contd....

Table 4.5 contd.

h	a_p^+	a_m^-	n^*	div
41	556	–90	108.1385	108.1393
42	93	–510	110.8453	110.8458
43	146	–181	113.5524	113.5535
44	318	–112	116.2597	116.2605
45	87	–2573	118.9672	118.9673
46	128	–266	121.675	121.6751
47	231	–144	124.383	124.384
48	989	–100	127.0912	127.0918
49	115	–460	129.7996	129.8
50	185	–192	132.5082	132.5093
51	442	–123	135.217	135.2177
52	106	–1319	137.926	137.9256
53	157	–274	140.6351	140.6357
54	295	–156	143.3444	143.3459
55	1918	–110	146.0539	146.0542
56	138	–446	148.7635	148.7637
57	226	–204	151.4733	151.4744
58	595	–134	154.1832	154.1838
59	124	–1039	156.8933	156.8934
60	187	–285	159.6035	159.6038
61	365	–168	162.3138	162.3152
62	4815	–120	165.0242	165.0243
63	161	–447	167.7348	167.7352
64	270	–218	170.4455	170.4467
65	782	–146	173.1563	173.1573
66	143	–935	175.8672	175.8673
67	218	–299	178.5783	178.5783
68	442	–181	181.2894	181.2905
69	228510	–131	184.0006	184.0006

Table 4.5 contd....

Table 4.5 contd.

h	a_p^+	a_m^-	n^*	div
70	185	–457	186.712	186.7118
71	315	–232	189.4234	189.4241
72	1004	–157	192.1349	192.1352
73	162	–895	194.8465	194.8467
74	249	–316	197.5582	197.5593
75	522	–194	200.27	200.271
76	146	–7862	202.9819	202.9818
77	209	–474	205.6938	205.694
78	361	–248	208.4059	208.4072
79	1259	–169	211.118	211.1183
80	181	–886	213.8302	213.8304
81	281	–334	216.5424	216.5431
82	605	–208	219.2548	219.2558
83	162	–4747	221.9672	221.967
84	233	–494	224.6797	224.6795
85	408	–264	227.3922	227.3929
86	1544	–181	230.1048	230.1049
87	200	–897	232.8175	232.8177
88	312	–353	235.5302	235.5308
89	689	–222	238.243	238.2437
90	177	–3831	240.9559	240.9558
91	256	–517	243.6688	243.6688
92	453	–281	246.3818	246.3828
93	1846	–194	249.0948	249.0951
94	219	–921	251.8079	251.8079
95	343	–374	254.521	254.5216
96	771	–237	257.2342	257.2351
97	193	–3468	259.9475	259.9473
98	279	–544	262.6607	262.661
99	498	–298	265.3741	265.3744

Table 4.5 contd....

Table 4.5 contd.

h	a_p^+	a_m^-	n^*	div
100	2152	–207	268.0875	268.0877
101	237	–955	270.8009	270.8012
102	373	–396	273.5144	273.515
103	850	–252	276.2279	276.2287
104	208	–3342	278.9415	278.9414
105	302	–573	281.6551	281.6548
106	541	–317	284.3688	284.3695
107	2441	–220	287.0825	287.0827
108	255	–998	289.7963	289.7965
109	402	–419	292.51	292.5103
110	925	–267	295.2239	295.224
111	223	–3352	297.9377	297.9376
112	323	–606	300.6516	300.6523
113	582	–336	303.3656	303.366
114	2696	–234	306.0796	306.0799
115	273	–1049	308.7936	308.7935
116	430	–444	311.5077	311.508
117	993	–284	314.2218	314.2224
118	237	–3461	316.9359	316.9359
119	345	–641	319.6501	319.6501
120	620	–356	322.3643	322.3647
121	2902	–248	325.0785	325.0787
122	290	–1108	327.7928	327.7926
123	457	–470	330.5071	330.507
124	1054	–300	333.2214	333.2216
125	251	–3658	335.9357	335.9358
126	365	–678	338.6501	338.6501
127	656	–377	341.3646	341.365
128	3050	–262	344.079	344.0791
129	306	–1176	346.7935	346.7935
130	482	–498	349.508	349.5081

Table 4.6 Clifford subalgebras in pole n^* of φ-factor for malignant tumour.

h	**Main Clifford algebra** ***Cl*(. , .)**	**Clifford subalgebras** ***Cl*(. , .)**				
8	78, 19	13, 9	7, 6	5, 5	4, 4	3, 3
18	7046, 35	34, 18	17, 12	11, 9	9, 8	7, 6
25	53, 390	25, 43	16, 23	12, 16	10, 12	8, 10
28	54, 2824	27, 52	18, 27	13, 18	11, 14	9, 11
38	3603, 74	70, 37	35, 25	24, 19	18, 15	14, 13
45	87, 2573	43, 83	29, 42	21, 29	17, 22	14, 18
50	185, 192	62, 64	38, 39	27, 28	21, 22	17, 18
62	4815, 120	114, 60	58, 40	39, 30	29, 24	23, 20
69	228510, 131	130, 66	65, 44	43, 33	33, 27	26, 22
76	146, 7862	72, 141	48, 72	36, 48	29, 36	24, 29
83	162, 4747	79, 152	53, 78	39, 52	32, 40	26, 32
114	2696, 234	199,113	103, 74	70, 56	53, 44	42, 37
123	457, 470	154,156	92, 94	66, 67	51, 52	42, 43
125	251, 3658	122,222	80, 115	60, 78	48, 59	40, 47
128	3050, 262	223,126	116, 83	78, 62	59, 50	48, 42
130	482, 498	162,165	98, 99	70, 71	54, 55	45, 45

THE FUTURE OF ONCOLOGY

The study of oncoviruses has led to the understanding that oncogenes play a very great (if not critical) role in origin of malignant tumours. Then, it became clear that genes-suppressors are opposed to oncogenes.

We should recall such substances as ferroelectrics and ferromagnetics with effect of spontaneity. May be cancers have some analogy with them? Gigantic entanglement of malignant tumours with very great φ-factor arises in many cases spontaneously because Brownian particles are always available in the cell. What is common between exchange energy of spins and stochastic exchange of states in attractor? We discovered that malignant tumours have gigantic φ-factor. A conclusion comes to mind that there should be a chain leading to "cancer":

oncogene → attractor → fractal.

Passing from cell level to the level of intracellular structures, the "problem of cancer" does not become simpler, since with decreasing sizes of biological structures the structure of fractal space of states remains rather complicated.

We believe that further progress of theoretical oncology is connected with developing conception of attractor proteins, which may change the value of *h* parameter of the cell and decrease its φ-factor. Now we cannot measure φ-factor of cell—living cell is considered infinitely complex. Choosing Brownian process as etalon, we may hope to determine how cell processes are more complex and entangled than the Brownian process. Possibly, only conception of φ-factor will be insufficient.

"Cancer problem" is complex because it has no physical analog. Let us recall the Bogolubov principle of correlation decaying: sooner or later higher correlations in molecular movement become simpler. However, "cancer" is an exclusion: higher correlations do not become simpler. New purposeful medicinal preparations are synthesized—but has anybody measured their φ-factor? If φ-factor > 1, such a drug may be an oncoinductor! Besides, infinite number of causes may lead to oncodiseases. Entanglement is the category that cannot be easily measured. For example—is the nucleus of an atom more entangled then its shell or not?

Old methods cannot cope with "oncoproblem". Linear-differential method for determining function space of genes exhausted itself long ago. As soon as we deal with genetic instability, φ-factor reveals itself at once. We believe that investigation of virus oncogenesis is the most correct way to an advanced understanding of "cancer" nature owing to minimal uncertainty of primary agent. We may attempt to reduce great φ-factor. We should search for substances or *natural microorganisms* that can lower φ-factor of cells. If φ-factor of the malignant cell was at the level 2…5, we would hope that two or three proteins could solve all the problems. But at φ-factor equal to 10000, it is practically impossible to look into such "onco-process"—the magic drug does not exist.

Nevertheless, we should not lose hope. In cells of benign tumours, φ-factor is negative. Using methods of cell technologies, we may try to "retarget" such cells with great negative φ-factor to form a capsule around malignant tumour.

Selected Bibliography

Stcherbic, V.V., Buchatsky, L.P. Generation of Henon attractor at oncogenic polyomavirus cell transformation. Scientific Transactions of Ternopil National Pedagogical University. Ser. Biol. 50: 110–119, 2012 (in Ukrainian).

Carcinogenesis

Epifanova, O.I. Lectures on cell cycle. Moscow: KMK Scientific Press Ltd., 2003 (In Russian).

The Transformed Cell. Edited by I.L. Cameron, T.B. Pul. New York: Academic Press, 1981.

Buchatsky, L.P., Galakhin, K.A. Tumours of fish in reservoirs of Ukraine. Kyiv: DIA, 2009 (In Russian).

Kopnin, B.P. Targets of oncogenes and tumour suppressors: A key for understanding basic mechanisms of carcinogenesis. Russian Journal of Biochemistry 65: 5–33, 2000 (In Russian).

Principles of molecular oncology. Edited by M.H. Bronchud, MaryAnn Foote, W.P. Peters, M.O. Robinson. Totowa, New Jersey: Humana Press, 2000.

The molecular basis of cancer. Editor: Dolores Meloni. Philadelphia: Saunders, 2008.

Cell cycle and growth control: Biomolecular regulation and cancer. Edited by Gary S. Stein, Arthur B. Pardee. Hoboken, New Jersey: John Wiley & Sons, Inc., 2004.

Michor, F., Yoh Iwasa and Nowak, M.A. Dynamics of cancer progression. Nature reviews/ cancer 4: 197–206, 2004.

Weber, G.F. Molecular Mechanisms of Cancer. Dordrecht, The Netherlands: Springer, 2007.

Fasman, G.D. The handbook of biochemistry and molecular biology. V. 1. CRC Press, 1975.

Marino-Ramirez, L., Hsu, B., Baxevanis, A.D., Landsman, D. The histone database: a comprehensive resource for histones and histone fold-conteining proteins. Proteins 62: 838–842, 2006.

Poliomaviruses

Caspar, D.L., Klug, A. Physical principles in the construction of regular viruses. Cold Spring Harbor Symp. Quant. Biol. 27: 1–24, 1962.

Fields virology. Edited by David M. Knipe, Peter M. Howley. Philadelphia: Lippincott Williams & Wilkins, 2001.

Rosenthal, L.J., Doerr, H.W. (Editors). Mechanisms of DNA Tumor Virus Transformation. Basel: Karger, 2001.

Stehle, T., Gamlin, S.J., Yan, Y., Harrison, S.C. The structure of simian virus 40 refined at 3.1 Å resolution. Curr. Biol. 4: 165–182, 1996.

Ben-nun-Shaul, O., Bronfeld, H., Reshef, D., Schuler, O., Oppengeim, A. The SV40 capsid is stabilized by a conserved pentapeptide hinge of the major capsid protein VP1. J. Mol. Biol. 386: 1382–1391, 2009.

Salunke, D.M., Caspar, D.L.D., Garsea, R.L. Polymorphism in the assembly of polyomavirus capsid protein VP1. Biophys. J. 56: 887–900, 1989.

Fanning, E., Zhao, K. SV40 DNA replication: from the *A* gene to nanomachine. Virology 384: 352–359, 2009.

Robles, M.T.S., Pipas, J.M. T-antigen transgenic mouse models. Semin Cancer Biol. 19: 229–235, 2009.

Gjoerup, O., Chang, Y. Update on human polyomaviruses and cancer. Adv. in Cancer Res. 1–51. Elsevier Inc., 2010.

Khalili, K., Sariyer, L.K., Safak, M. Small tumor antigen of polyomaviruses: role in life cycle and cell transformation. J. Cell Physiol. 215: 309–319, 2008.

Sullivan, C.S., Pipas, J.M. T antigens of Simian virus 40: molecular chaperones for viral replication and tumorigenesis. Microb. and Mol. Biol. Rev. 66: 179–202, 2002.

Noble, J.C.S., Prives, C., Manley, J.L. Alternative splicing of SV40 early pre-mRNA is determined by branch site selection. Genes & Dev. 2: 1460–1475, 1988.

Butenko, Z.A., Phylchenkov, A.A. Modern concepts of viral oncogenesis: Fundamental and applied aspects. Exp. Oncol. 22: 239–245, 2000 (in Russian).

Attractors

Henon, M. A two-dimensional mapping with a strange attractor. Commun. Math. Phys. 50: 69–77, 1976.

Nikolis, G., Prigogine, I. Exploring complexity. An introduction. New York: W.H. Freemann and Company, 1989.

Zaslavsky, G.M., Sagdeev, R.Z., Usikov, D.A., Chernikov, A.A. Weak chaos and quasi-regular patterns. Cambridge: Cambridge University Press, 1991.

Lui Lam. Introduction to nonlinear physics. New York: Springer-Verlag, 1997.

Fractals

Crownover, R.M. Introduction to fractals and chaos. Boston: Jones and Bartlett Pub., 1995.

Zaslavsky, G.M., Sagdeev, R.Z. Introduction to nonlinear physics: From pendulum to turbulence and chaos. Moscow: Nauka, 1988 (in Russian).

Falconer, K. Fractal geometry: Mathematical foundations and applications. New York: John Wiley & Sons, 1990.

Feder, J. Fractals. New York: Plenum Press, 1988.

Kaandorp, J.A. Fractal modeling. Growth and form in biology. Berlin: Springer-Verlag, 1994.

Dewey, G.T. Fractals in molecular biophysics. Oxford: Oxford University Press, 1997.

Janecka, I.P. Cancer control through principles of systems science, complexity and chaos theory: A model. Int. J. Med. Sci. 4: 164–173, 2007.

Mandelbrot, B.B. The fractal geometry of nature. New York: W.H. Freeman and Company, 1977.

Synergetics and nanotechnologies

Hermann Haken. Synergetics. An introduction. Nonequilibrium phase transitions and self-organization in physics, chemistry and biology. Berlin: Springer-Verlag, 1977.

Riabykh, T.P., Osipova, T.V., Sokolova, Z.A. et al. Protein nano- and microparticles for diagnostics of oncologic diseases.

Nanotechnological society of Russia. http://ntsr.info/science/library/2972.htm (in Russian).

Boas, U., Christensen, J.B., Heegaard, P.M.H. Dendrimers in Medicine and Biotechnology. New Molecular Tools. Cambridge: The Royal Society of Chemistry, 2006.

Lectures and visual materials of Science and educational Centre of MSU on nanotechnologies. http://www.nanometer.ru/lectures.html (in Russian).

Cancer Nanotechnology. Methods and Protocols. Edited by Stephen R. Grobmyer and Brij M. Moudgil. New York: Humana Press, 2010.

5

Transformation of Genetic Information into Clifford Algebra

Primary structures of RNA and protein molecules are long polymer chains. Functional properties of these molecules become discernible only after their transformation into three-dimensional structures, which frequently contain modified elements. Folding of RNA and protein molecules is a characteristic feature of biological processes.

From the standpoint of transformation of biomolecule into Clifford algebra, we will consider in this chapter RNA splicing and formation of discrete structure of protein molecules. In Chapter 1, initial RNA sequence was transformed to Clifford algebra $Cl(5, 4)$. However, such algebra has small dimension, equal to 512. To realize Clifford algebra, the number of nucleotides should be about one third of the algebra dimension. To realize Clifford algebra $Cl(15)$, it would require approximately $2^{15}/3 \approx 109000$ RNA nucleotides. RNA is clearly difficult to realise Clifford algebras with higher dimensions. Here we propose a hypothesis that **Clifford algebra of the order > 10 may be realized by protein molecules.**

Linear sequences of RNA nucleotides as well as amino acid sequences of proteins do not generate Clifford algebra, as there is no pairing of coding elements. Primary sequence of RNA nucleotides consists of intervals [3+ 1–], which may be presented as **A** = [+ + + –], **U** = [+ + – +], **C** = [+ – + +], **G** = [– + + +]. It is impossible to build Clifford algebra only with intervals [3+ 1–]. Clifford algebra in RNA molecule is expanded into algebra $Cl(3)$. To realise Clifford algebras of greater dimensions, biological structures, which allow expansion in algebra of the order no less then $Cl(10)$ order, are necessary. We will show that just Clifford algebra $Cl(10)$ is algebra of expansion of Clifford algebras in protein molecules.

Generation of Clifford algebra of greater dimensions on RNA is also possible at the expense of nucleotide modification. If we assume that up to 100 modified nucleotides may be presented in different tRNA, then Clifford algebra in tRNA molecule may be expended into algebra $Cl(6)$.

RNA SPLICING

In Chapter 1, formulae were given for calculating numbers of nucleotides $n_1 \ldots n_8$ that determine secondary RNA structure as some Clifford algebra:

$$\{3(\mathbf{P+}) - (\mathbf{Q-})\}/4 = 2n_{765} + 3n_8 + n_{432};$$
$$\{3(\mathbf{Q-}) - (\mathbf{P+})\}/4 = 2n_1 + n_{432} - n_8,$$

where $n_{765} = n_7 + n_6 + n_5$, $n_{432} = n_4 + n_3 + n_2$. It follows from these formulae that the sum of single and paired RNA nucleotides is equal to one fourth of Clifford algebra dimension. Therefore, realisation of the same Clifford algebra is possible within wide limits of total number of RNA nucleotides. We may suppose that the change of RNA nucleotide number at splicing is connected with different realisation of Clifford algebra. Besides, splicing is essential, as primary RNA transcript may not generate Clifford algebra.

RNA splicing may also be associated with partial decrease of the volume of vector space of Clifford algebra—real RNA do not contain all obligatory vectors of the algebra.

Collective numbers n_{765} and n_{432} allow wide discretion about choosing sequences of RNA nucleotide. This situation is similar to choosing matrices for generating elements of Clifford algebra in affine space.

Splicing phenomenon

RNA and protein splicings are the most important biochemical processes underlying preservation and transformation of genetic information. RNA splicing indicates a mosaic structure of genes and demonstrates partial independence of RNA information on initial sequence of genes transcribed on DNA matrix.

RNA splicing, i.e., cutting out of RNA fragments (introns) and linkage of remaining RNA parts (exons), takes place after DNA transcription. Most common are two large groups of introns: group I introns with linear structure and group II introns forming lariat like structures. Introns of groups I and II possess autocatalytic activity, that is, they are ribozymes.

Introns of group II are cut out in pre-mRNA by spliceosome. Splicing of group I introns is characterised by presence of outer guanosine cofactor. Besides, group I introns may perform the function of nucleotidyl-transferase and endoribonuclease. Cofactors are not required cut out introns of group II; splicing is initiated by internal adenine of the intron. There are also alternative splicing, trans-splicing and tRNA splicing. Besides RNA splicing, there is also protein splicing.

Biochemistry of splicing is very complicated. Splicing requires availability of specified 3′- and 5′-sequences. Splicing is catalyzed by a large complex containing RNA and proteins (spliceosome). Spliceosome includes five small nuclear ribonucleoproteins (snRNA)—U1, U2, U4, U5 and U6. RNA, in composition of snRNA, interacts with intron and, possibly, participates in catalysis. It takes part in splicing of introns, which contain **GU** in 5′-site and **AG** in 3′-site. Sometimes during the process of maturation, mRNA may undergo alternative splicing, when introns in content of pre-mRNA may be cut out in different alternative combinations; at that, exons may also be frequently cut out. As a result of alternative splicing of one gene pre-mRNA, numerous mRNA and their protein products may be created.

Conception of enzymes changed after discovery of RNA self-splicing: besides proteins, RNAs also possess enzymatic activity. Even more surprising is alternative splicing, when different protein molecules may be obtained from one gene. Alternative splicing is a characteristic feature of RNA processing in eukaryotes. Sometimes, trans-splicing introns may be of an alternative type, when a point of adenine site produces RNA branching.

Functional differences between RNA and proteins became still less, when in 1990 the group of T. Stevens discovered the phenomenon of protein autosplicing.

Group of alternatives

Different types of splicing possess high accuracy of intron cutting. Alternative splicing, performed by spliceosomes, resembles deterministic chaos: on the one hand, probabilistic nature of alternatives is present, on the other—high accuracy of RNA splicing. It may be supposed that the alternative is governed by the symmetry of splicing process—a continuous group.

Classical representations of groups (functional) are frequently used in quantum mechanics. Vectors of states are connected between themselves by the group symmetry. Measuring of a physical process result in quantum mechanics nearly always strongly diminishes the role of the symmetry of states. However, in biological systems, enzymes are active molecules. It may be supposed that they do not diminish but increase the role of symmetry between the states of a biochemical processes. This results in high accuracy of genetic information transformation, which is characteristic of splicing phenomenon.

Group of alternatives (well-known group ISO(2)) do not describe the splicing process, but allows the interpretation of intron structures. The rule **GU–AG** for group II introns, as well as the presence of spliceosome, apparently, cannot be understood with the help of the group ISO(2). We also miss explanation of capping and polyadenylation of pre-mRNA. Besides, we should accentuate at once that further constructed representations of the group ISO(2) will be named ***alternative***, as they contain many nontrivial vector interpretations.

Group ISO(2) is a very simple group, which is related to photon spin. It is a group of shifts and rotations of Euclidean plane that is well studied. We use vector interpretation of the group ISO(2). Vectors of the group are RNA or protein fragments. We are also using parallel transfer of vectors at their addition.

Group ISO(2) is a group of second order complex matrices of the following structure:

$$G(\alpha, r, \varphi) = \begin{vmatrix} \exp i\alpha & r\exp i\varphi \\ 0 & 1 \end{vmatrix}$$

where

$r \exp i\varphi$ is radius-vector of Euclidean plane point;
$\exp i\alpha$—phase factor of the same point after shifting along OX axis.

Group ISO(2) is a Lie group. It is not difficult to calculate group generators, which satisfy the following commutation relations:

$$[a_1, a_2] = 0; [a_2, a_3] = a_1; [a_3, a_1] = a_2.$$

In our reasoning we assign a great role to inner automorphisms of ISO(2) group, which are generated by the group action on itself through transformation of similarity:

$$G(\alpha_0, r_0, \varphi_0) \to G(\alpha, r, \varphi)\, G(\alpha_0, r_0, \varphi_0)\, G(\alpha, r, \varphi)^{-1},$$

that is, instead of one vector of ISO(2) group, three vectors are used. Group commutators are also used.

Equation of alternative will be recorded as automorphism of ISO(2) group in the form $g_s = gg_0g^{-1}$, where

$$g = \begin{vmatrix} \exp i\alpha & r\exp i\varphi \\ 0 & 1 \end{vmatrix};\quad g^{-1} = \begin{vmatrix} \exp(-i\alpha) & -r\exp i(\varphi-\alpha) \\ 0 & 1 \end{vmatrix};$$

$$g_0 = \begin{vmatrix} \exp i\alpha_0 & r_0\exp i\varphi_0 \\ 0 & 1 \end{vmatrix};$$

$$g_s = \begin{vmatrix} \exp i\alpha_0 & -r\exp i(\varphi+\alpha_0) + r_0\exp i(\varphi_0+\alpha) + r\exp i\varphi \\ 0 & 1 \end{vmatrix}.$$

Figure 5.1 shows diagrams of ISO(2) group alternatives.

Vectors of the direct alternative are linearly independent, while vectors of the reversed alternative are dependent on arrangement of direct alternative vectors.

At addition of alternative vectors into one vector, their reference point is supposed to remain fixed. Orientation of any vector may be changed by shifting its phase reference line by an angle equal to 180°.

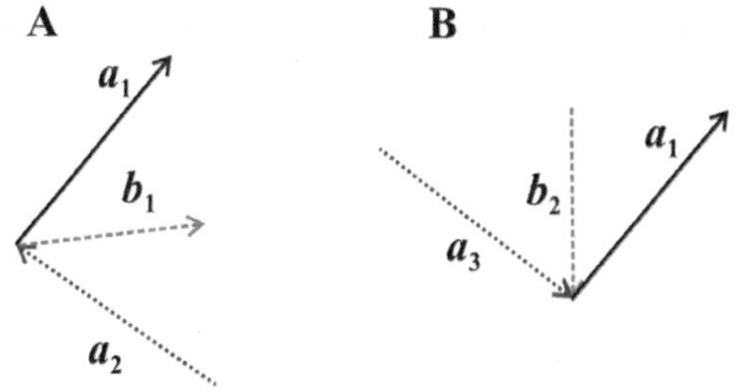

Figure 5.1 Alternatives of ISO(2) group.
(A) Direct alternative g_s. **(B)** Reversed alternative g_s^{-1}.
$a_1 = +r\exp i\varphi$, $a_2 = -r\exp i(\varphi + \alpha_0)$, $a_3 = -r\exp i(\varphi - \alpha_0)$,
$b_1 = +r_0\exp i(\varphi_0 + \alpha)$, $b_2 = -r_0\exp i(\varphi_0 + \alpha - \alpha_0)$.
Numerical values of angles are chosen as: $\varphi_0 = -\alpha$, $\varphi = 45°$, $\alpha_0 = -90°$.

Diagrams of introns

It is clear that, if vectors of alternative are mechanically added into one vector, we may obtain only linear intron of group I. That is why parallel transfer of vectors and their relative position play a key role at plotting intron diagrams.

Figure 5.2 shows main intron diagrams at splicing of RNA and proteins. Let us describe each diagram of introns in detail.

Diagram <1>. Linear introns of group I. Arbitrary number of ISO(2) group vectors are added in order to obtain these introns.

Diagram <2>. This diagram also relates to group I introns. Outer guanine **G** attacks intron and linear introns are formed as a result of two transesterification reactions. Vectors $\boldsymbol{b}_1$ and $\boldsymbol{a}_2$ are guanine lines of force. Vectors $\boldsymbol{a}_1$ is intron. Here, ISO(2) group is not homogeneous.

Diagram <3>. This is the main diagram of group II introns. As a result of closure of RNA from site **A**, lariat figure is formed. Introns in the form of lariat may also be created at pre-RNA processing with spliceosome and at alternative splicing. Instead of three vectors, their greater number may be used: order of vector summation in the lariat loop is of no importance.

Diagram <4>. Group I intron closed into a circle. The circle may quickly disintegrate or remain for quite a long time.

Diagram <5>. Direct intron alternative that may be observed at trans-splicing. Surprisingly, Y-intron, which corresponds to ISO(2) group automorphism, does not correspond to RNA self-splicing.

Diagram <6>. Protein splicing. Amino acid residues **Asn** and **Gln** are modulus of protein backbone. Vector $\boldsymbol{a}_2^+$ of sequence of amino acid residues in protein structure is a continuation of vector $\boldsymbol{a}_1^+$ of periodical protein

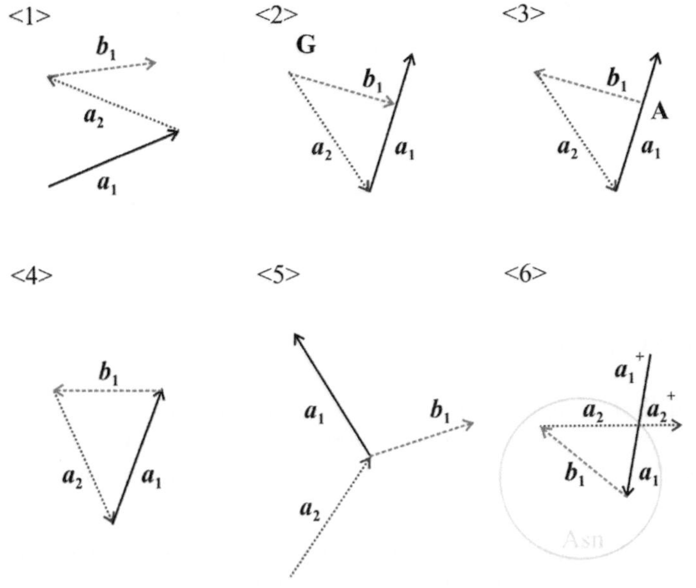

Figure 5.2 Main intron diagrams at splicing of RNA and proteins.

backbone. Since a periodical backbone sequence has natural orderliness, this splicing diagram supposes availability in proteins of ordered periodical fragments—α-helices and β-sheet. Here ISO(2) group is homogeneous.

Alternative splicing

Because of strong mutual dependence of vectors, application of ISO(2) group automorphism for interpreting of alternative splicing is ineffective. Therefore, we will consider group transformation of vectors not on the basis of automorphism but in terms of recurrent transformation of introns (*embedding of introns into commutators of the group*). Each intron is situated between exons, and group inverse elements of the last are associated with intron.

Let us consider the simplest case, when 1 or 2 introns are cut out from primary RNA transcript (Figure 5.3).

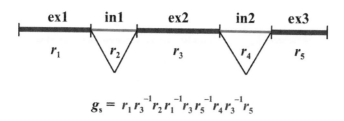

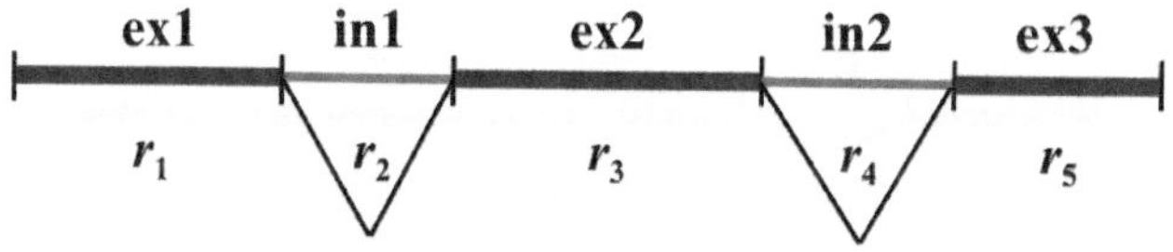

$$g_s = r_1 r_3^{-1} r_2 r_1^{-1} r_3 r_5^{-1} r_4 r_3^{-1} r_5$$

Figure 5.3 Alternative splicing of two introns.
Exons: ex1, ex2, ex3; introns: in1, in2.
Cut out: in1, in2 or only in1.
g_s—group element of splicing.

Each cutting out intron is surrounded by two reverse exons in reverse order. In diagrams of splicing (Figure 5.4), exon vectors remain at the beginning of reading, while intron vectors form a lariat.

Vectors r_1^{-1}, r_3^{-1}, r_5^{-1} are fictitious; in sum, they show the path of lariat in intron or rotation vector in exon. Vector $r_4^* = r_4 + r_3^{-1} + r_5^{-1}$.

Alternative splicing with partial exon cutting out is shown in Figure 5.5.

Group element is written in two variants, as two inverse elements of ISO(2) group are sufficient for cutting out one intron.

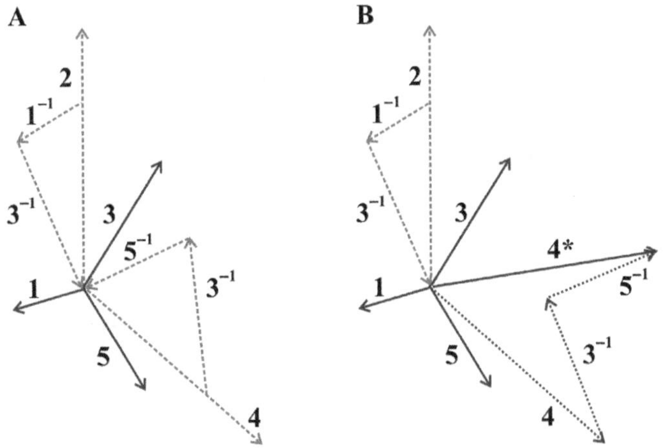

Figure 5.4 Diagrams of alternative splicing of two introns. (**A**) Cut out: in1 and in2. (**B**) Cut out: in1.

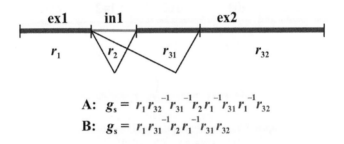

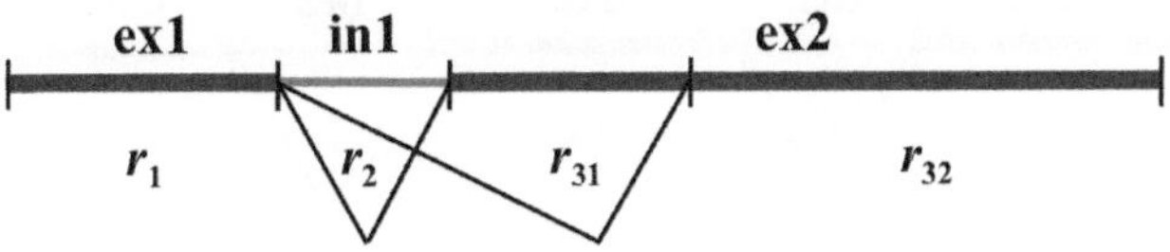

$$\text{A: } g_s = r_1 r_{32}^{-1} r_{31}^{-1} r_2 r_1^{-1} r_{31} r_1^{-1} r_{32}$$
$$\text{B: } g_s = r_1 r_{31}^{-1} r_2 r_1^{-1} r_{31} r_{32}$$

Figure 5.5 Alternative splicing with partial exon cutting out. In exon $(r_{31} + r_{32})$, part r_{31} is cut out. (**A**) complete version. (**B**) condensed version.

Figure 5.6 shows diagrams of condensed version of alternative splicing with partial cutting out of exon.

Alternative splicing of RNA of SV40 virus is given in Figure 5.7. Here, small T-antigen, equal to the sum of exon ex1 and intron in1, terminates inside intron (in1 + in2); therefore, additional fictitious vector r_5 should be introduced for correct organization of commutator.

Figure 5.8 gives diagrams of alternative splicing of SV40 T-antigens.

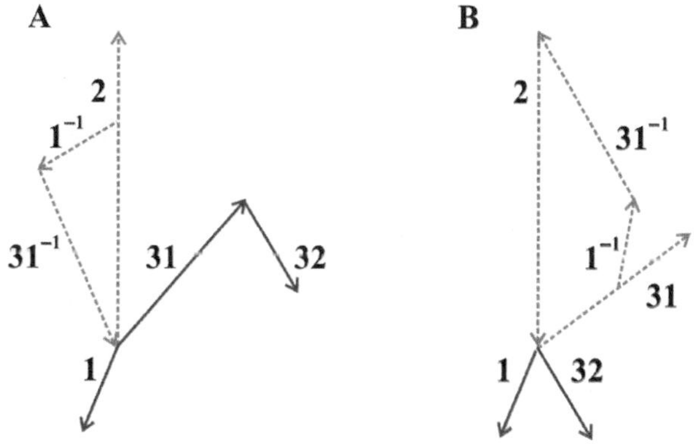

Figure 5.6 Diagrams of alternative splicing with partial cutting out of exon. (**A**) Cut out: in1. (**B**) Cut out: in1 and r_{31}.

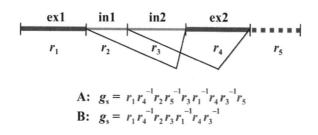

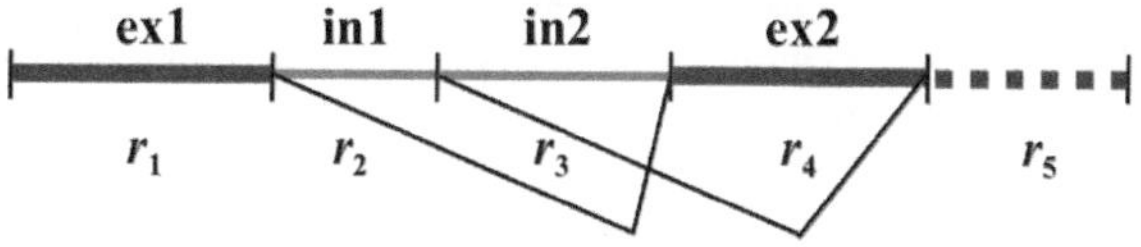

$$\text{A:}\quad g_s = r_1 r_4^{-1} r_2 r_5^{-1} r_3 r_1^{-1} r_4 r_3^{-1} r_5$$
$$\text{B:}\quad g_s = r_1 r_4^{-1} r_2 r_3 r_1^{-1} r_4 r_3^{-1}$$

Figure 5.7 Alternative mRNA splicing of large (ex1 + ex2) and small (ex1 + in1) T-antigens of virus SV40. (**A**) complete version. (**B**) condensed version ($r_5 = 1$).

Generally, at alternative splicing, alternative diagrams of introns are present. Here, ISO(2) group is not homogeneous. Besides, summation of exon vectors is performed in sequence collinear to sequence of exons in pre-mRNA, since ISO(2) group does not suppose any orderliness at vector summation.

It is clear that ISO(2) group in its vector interpretation is a good fit for describing different types of introns at RNA splicing and inteins at protein splicing. However, at description of alternative splicing with the use of ISO(2) group, we have to apply commutators, as group automorphisms are ineffective in this case.

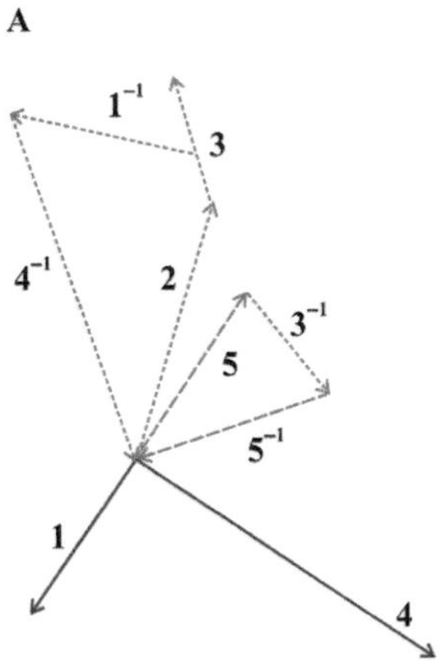

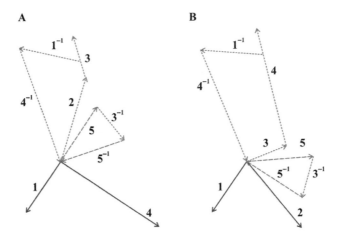

Figure 5.8 Diagrams of alternative splicing of SV40 T-antigens with partial cutting out of exon (complete version).
Vector $\mathbf{r}_5$ is located in fictitious circular intron. In condensed version, reversed vectors $\mathbf{r}_1^{-1}$, $\mathbf{r}_4^{-1}$, $\mathbf{r}_3^{-1}$ are joined sequentially.
(**A**) Cut out: in1 and in2. (**B**) Cut out: in2 and $\mathbf{r}_4$.

DISCRETE SYMMETRY OF GLOBULAR PROTEINS

Globular proteins are crucial components of cell structures. Curving in space in a most complicated way, protein polypeptide chain forms a compact molecule. At atomic level, folding of polypeptide chain always results in asymmetric structure; outwardly, globular protein is a rounded irregular-shaped particle. Globular proteins do not have any discrete symmetry as in regular polyhedrons.

A multitude of experimental data based on X-ray structure analysis (PDB database) permit us to state with great certainty that globular proteins at the level of secondary structure have many common components: relatively short regions of α -, 3_{10} – helices and β-sheets, few **S-S** bonds and single contacts between different parts of polypeptide chain formed by hydrogen bonds. Globular proteins contain few non paired amino acid residues.

At the level of tertiary structure, there are ion bonds between oppositely charged side groups of amino acid residues and hydrogen bonds between side groups and backbone of the chain, which occurs very seldom. There are electrostatic interactions between atoms of protein globule, hydrophilic and hydrophobic interactions with surrounding water molecules as well as

Van der Waals forces. Hydrogen bonds and Van der Waals forces perform the leading role in stabilization of globular proteins.

In native state, protein molecules have very high packing factor, therefore, protein globules are similar to the crystals of small organic molecules. Nevertheless, crystal symmetry in proteins is not observed.

In globular proteins, polypeptide chain is folded in such a way that each chain link is oriented relative to its neighbours differently, remaining in the limits of allowed conformational angles φ and ψ. There is relationship between the building of stable secondary structures (helices and β-sheets) and values of conformational angles (Ramachandran plots).

Key moment, determining correct functioning of globular proteins, is the formation of their surface. Proteins with different surface configuration usually perform different functions. This configuration does not have any clearly defined symmetry elements.

Globular proteins are modular structures with regularly repeated monomeric units. Therefore, we may hope to distinguish modular symmetry in the structure of globular proteins. We should note at once that crystal symmetry is impossible for protein globule because of polypeptide chain finitude. On the basis of experimental data, we will show that globular proteins have discrete modular symmetry, which is described by the group SL (2, 5).

Modular group SL(2, 5)

To find generators of discrete symmetry group of protein molecule, we should select obviously discrete elements of polypeptide chain secondary structure. First of all, these are hydrogen bonds between **NH**- and **CO**-groups of backbone, as well as cysteine **S-S** bridges. To allocate precisely these elements, we used the *DSSP*-program, one of the most widespread programs that define secondary structure. The DSSP algorithm defines a hydrogen bond when the bond energy is below –0.5 kcal/mol from an electrostatic approximation of the hydrogen bond energy.

We identify each hydrogen bond or **S-S** bridge as amino acid pairing. Based on pairing, we construct two quadratic forms of globular protein. These forms are rather far from standard geometric quadratic forms of the surface. First quadratic form is defined as pairing of coordinates of amino acid residues on basis of hydrogen bonds or **S-S** bridges. Second quadratic form (alphabetic-literal) is defined on pairing amino acid residues as elements of alphabet from twenty letters.

Let us consider in detail the constructing of quadratic form for 1XPT protein, as an example. *DSSP*-program allows determining coordinates of paired amino acid residues.

To construct quadratic form, the form coefficients and their signs are needed. For separate pair $h_i h_j$ of first quadratic form, we have chosen the following coefficients c_{ij} and their signs z_{ij}:

$c_{ij} = 2$, $z_{ij} = +1$, if amino acid residues *i* and *j* are bound by hydrogen bond;

$c_{ij} = 2$, $z_{ij} = -1$, if amino acid residues *i* and *j* are bound by **S-S** bridge;

$c_{ii} = 1$, $z_{ii} = +1$, if amino acid residue *i* is proline.

Sign z_{ij} takes account also of parity of amino acid arrangement: $z_{ij} \to z_{ij}(-1)^{i+j}$.

For 1XPT protein, first quadratic form F1 is given by:

F1(1XPT) =

+2h7h3 +2h8h4 +2h9h5 +2h10h6 +2h11h7 +2h12h8 +2h13h9 –2h17h14 –2h25h22 +2h28h24 +2h29h25 +2h30h26 +2h31h27 +2h32h28 +2h33h29 –2h34h31 –2h35h30 –2h37h34 +2h41h35 +2h41h39 +1h42h42 –2h47h12 –2h47h14 +2h49h47 +2h54h50 +2h55h51 +2h56h52 –2h57h54 –2h58h55 –2h59h56 –2h60h57 +2h62h60 –2h68h65 + 2h69h65 –2h72h63 –2h72h63 –2h74h61 –2h74h61 +2h80h48 +2h81h79 +2h82h46 + 2h82h46 +2h84h44 +2h84h44 +2h86h42 –2h90h87 +1h93h93 –2h94h91 –2h94h91 +2h95h93 –2h96h87 –2h98h85 –2h98h85 –2h100h83 –2h100h83 –2h102h81–2h102h81 –2h104h79 –2h104h79 –2h106h75 –2h108h73 –2h108h73 –2h110h71 +1h114h114 –2h116h111 –2h116h111 +2h116h114 +1h117h117 –2h118h109 +2h119h109 +2h119h109 +2h121h107 +2h121h119 –2h122h107 –2h122h107 –2h124h105 –2h84h26 +2h95h40 – 2h110h58 +2h72h65.

For separate pair $A_i A_j$ of second quadratic form, we chose the following coefficient c_{ij} and their signs z_{ij}:

$c_{ij} = 2$, $z_{ij} = +1$, if amino acid residues A_i and A_j are bound by hydrogen bond and are different;

$c_{ij} = 1$, $z_{ij} = +1$, if amino acid residues A_i and A_j are bound by hydrogen bond and coincide;

$c_{ij} = 1$, $z_{ij} = -1$, if amino acid residues A_i and A_j are cysteines and bound by **S-S** bridge;

$c_{ii} = 1$, $z_{ii} = +1$, if amino acid residue A_i is proline.

Sign z_{ij} takes account also of parity of amino acid arrangement: $z_{ij} \to z_{ij}(-1)^{i+j}$.

For 1XPT protein, second quadratic form F2 is given by:

F2(1XPT) =

+2KT +2FA +2EA +2RA +2QK +2HF +2ME −2TD −2YS +2QN +2MY +2MC +2KN + 2SQ +2RM −2NK −2LM −2KN +2KL +2KR +1PP −2VH −2VD +2EV +2VS +2QL +1AA −1VV −2CQ −2SA −2QV +2NQ −2GC +2QC −2CV −2CV −2QK −2QK +2SH +2IM +2TF +2TF +2CN +2CN +2EP −2ST +1PP −2NK −2NK +2CP −2AT −2KR −2KR −2TD −2TD −2AI −2AI −2KM −2KM −2IS −2VY −2VY −2CN +1PP −2VE −2VE +2VP +1PP −2VA +2HA +2HA +2DI +2DH −2AI −2AI −2VH −1CC +1CC −1CC +1CC.

Using Lagrange method, we will reduce quadratic forms F1 and F2 to the sum of squares. Let us denote signature of obtained quadratic forms by [*p*+ *q*−]. Signatures of quadratic forms for different proteins are given in Table 5.1.

Table 5.1 Signatures of quadratic forms F1 and F2.

Protein	**PDB code**	**Number of amino acids**	**Form F1 signature**	**Form F2 signature**
Bovine ribonuclease A	1XPT	124	42+ 39−	10+ 9−
Human gamma-D crystallin	1HK0	173	57+ 53−	11+ 9−
Yeast phosphoglycerate mutase-3PG	1QHF	240	93+ 84−	10+ 9−
Human gamma-B crystallin	2JDF	175	62+ 55−	11+ 9−
Human insulin monomer	2JV1	52	20+ 19−	8+ 7−
Mutant of Human gamma-D crystallin	2KFB	174	56+ 52−	10+ 10−
Bovine beta-trypsin	3AAV	223	80+ 75−	11+ 9−

Note: auxiliary amino acid, taking account of A1 chain discontinuity, is introduced for insulin 2JV1.

Additional element, which is a generator of globular protein discrete symmetry, is 4-code for directions of φ and ψ angle rotation.

For each amino acid, there may be introduced 4-code for rotation χ_4 of plane of peptide bond round angles φ and ψ according to the formula:

$$\gamma_4(\mathrm{Am}) = \begin{cases} 0,\ 1 \text{ - if } \varphi < 0;\ \psi < 0,\ \psi > 0 \\ 2,\ 3 \text{ - if } \varphi > 0;\ \psi < 0,\ \psi > 0 \end{cases}$$

Suppose that the first amino acid is the high-order bit of 4-code. Then we may obtain numerical value Γ_4 of 4-code of rotation of peptide bond planes relatively angles φ and ψ.

For 1XPT protein, 4-code χ_4 is given by (cysteine pairs are denoted by capital letters):

```
KETAAAKFERQHMDSSTSAASSSNYaNQMMKSRNLTKDRbKPVNTFVHESLADVQAVcSQKN
31100000000011001011010100000000130100111011111111000000001011

VAdKNGQTNdYQSYSTMSITDaRETGSSKYPNbAYKTTQANKHIIVAcEGNPYVPVHFDASV
11101211111110111111111110011111111111111111111211111111110111
```

and its numerical value is $\Gamma_4 = 3.745715861864131 \times 10^{74}$.

It follows from Ramachandran plots that main configurations of φ and ψ angles are located in left half-plane; therefore, 4-code is rather close to binary code.

Suppose that the generators of discrete symmetry of globular proteins are two independent integer values: value of 4-code of polypeptide chain Γ_4 and ratio of two quadratic forms F1 and F2. According to Cantor, transformation of geometric form F1 in alphabetic-literal form F2 requires functions $\Psi = \{r\ (F2)\}^{\{r\ (F1)\}}$, where r (F) is maximal value of quadratic form signature index (p or q). This formula is the covering mapping of positive (negative) part of F1 form signature into positive (negative) part of F2 form signature. Table 5.2 gives Γ_4 and Ψ statistics for different proteins.

It follows from Table 5.2 that mean value of ratio {1}/{2} is equal to $1.70129 \approx 5/3$.

Table 5.2 Statistics of Γ_4 and Ψ generators.

PDB code	Number of Ψ functions	$LOG_{10}\Psi$ {2}	Value of 4-code Γ_4	$LOG_{10}\Gamma_4$ {1}	Ratio {1}/{2}
1XPT	10^{42}	42	3.7457×10^{74}	74.57353	1.77556
1HK0	11^{57}	59.35938	1.1944×10^{104}	104.07716	1.75339
1QHF	10^{93}	93	2.6014×10^{144}	144.41521	1.55285
2JDF	11^{62}	64.56634	1.9111×10^{105}	105.28128	1.63059
2JV1	8^{20}	18.06179	1.0141×10^{31}	31.00609	1.71666
2KFB	10^{56}	56	4.7780×10^{104}	104.67924	1.86927
3AAV	11^{80}	83.31141	1.5639×10^{134}	134.19420	1.61075

Icosahedron group I = A5 has ratio of generator powers 5/3 and is defined by the following relations between generators: $\{S\}^3 = \{U\}^5 = \{SU\}^2 = 1$. However, it is clear that the group icosahedron is not a group of globular protein symmetry; is only the divisor of the main group. Complete group of icosahedron symmetry I_h cannot be the main group because it contains the reflection. Therefore, minimal suitable modular group of globular protein symmetry is the group 2I = SL(2, 5).

Note, that icosahedron group A5 is not a subgroup of modular group SL(2, 5).

CLIFFORD ALGEBRA OF GLOBULAR PROTEINS

In RNA molecule, Clifford algebra is realized by genetic elements and their pairs. In globular proteins, such mode of realisation is impossible as dimension of the algebra is too high (for 1XPT protein, Clifford algebra of geometric form $Cl(42, 39)$ has dimension $2^{81} = 2.41785\times10^{24}$).

Alphabet-literal form is similar to algebra $Cl(10, 10)$, but algebra of expansion of Clifford algebra of geometric form is algebra $Cl(10)$. Indeed, 20 amino acid residues and their pairs define $20\times21 = 420$ elements of Clifford algebra. It should seem that for coverage of alphabet-literal form, algebra $Cl(9)$ with dimension 512 is sufficient. However, algebra $Cl(9) = Cl(6)\times Cl(3)$ disintegrates into internal algebra $Cl(6)$, which results in great number of neutral amino acid residues, and exterior hydrogen bond algebra $Cl(3)$ of decomposition of Clifford algebra of geometric form.

Clifford algebra $Cl(3)$ contains intervals different from Lorentz intervals, therefore, it is not suitable for decomposition of Clifford algebra of geometric form. We should select algebra $Cl(2)$, which realizes formation of external hydrogen bonds of protein molecule with water molecules and other molecular structures (beams of links of Lorentz intervals of decomposition of Clifford algebra of geometric form, which originate from elements of Clifford algebra of alphabet-literal form coverage). But then algebra $Cl(10) = Cl(8)\times Cl(2)$ with Clifford algebra of coverage of alphabet-literal form $Cl(8)$ will be algebra of decomposition of Clifford algebra of geometric form.

It is clear that coverage of Clifford algebra of geometric form by Lorentz intervals determines distances of propagation of beams of protein molecule influence on other biomolecules. The number of these intervals is very large although intervals themselves are very small (may be several angstrom).

If the number of amino acid residue pairs in protein molecule exceeds 236, it is an inevitable occurrence of neutral amino acid residues. Besides active centre, elements of Clifford algebra $Cl(8)$ are also present in other amino acid residues, which, for the most part, are components of pairs formed by hydrogen bond and **S-S** bridge. The nearer the protein molecule surface, the greater the probability that amino acid residue is an element of Clifford algebra. Algebra $Cl(8) = Cl(4)\times Cl(4)$ is the algebra of two fermions, and, possibly, of quasi-particle equivalent to hydrogen atom, as all amino acid residues and their pairs are continuation of glycine amino acid residue.

APPENDIX 5.1

CLIFFORD ALGEBRA OF SMALL T-ANTIGEN OF SV40 VIRUS

As an example, we will consider construction of Clifford algebra in RNA of small T-antigen of SV40 virus. Figure 5.9 shows nucleotide sequence and scheme of RNA nucleotide pairing (secondary structure).

```
AUGGAUAAAGUUUUAAACAGAGAGGAAUCUUUGCAGCUAAUGGACCUUCUAGGUCUUG
........((((((((..........((((((((.((...((((((...)))))))..
AAAGGAGUGCCUGGGGGAAUAUUCCUCUGAUGAGAAAGGCAUAUUUAAAAAAAUGCAA
..))..((((((((((((....)))))))........)))))))..........)))))
GGAGUUUCAUCCUGAUAAAGGAGGAGAUGAAGAAAAAAUGAAGAAAAUGAAUACUCUG
)))(((((.((((.....)))).))))))........................(((((
UACAAGAAAAUGGAAGAUGGAGUAAAAUAUGCUCAUCAACCUGACUUUGGAGGCUUCU
((.........((..((((.((((.....))))))))..))..((((...(((((.((
GGGAUGCAACUGAGGUAUUUGCUUCUUCCUUAAAUCCUGGUGUUGAUGCAAUGUACUG
(((((......(((((....)))))........)))))))((((..((((......))
CAAACAAUGGCCUGAGUGUGCAAAGAAAAUGUCUGCUAACUGCAUAUGCUUGCUGUGC
))))))..))))))))))..(((((((....((((...(((((((((((((((...((.(
UUACUGAGGAUGAAGCAUGAAAAUAGAAAAUUAUACAGGAAAGAUCCACUUGUGUGGG
((...))).)).))))))))...............(((....(((((((....))))))
UUGAUUGCUACUGCUUCGAUUGCUUUAGAAUGUGGUUUGGACUUGAUCUUUGUGAAGG
))..)))(((..((........))..)))..))))))).)))))....)))))))((((.
AACCUUACUUCUGUGGUGUGACAUAAUUGGACAAACUACCUACAGAGAUUUAAAGCUC
.......)))).(((((.((.((....))..)).)))))).)))))))).))))))))))).
```

Figure 5.9 Nucleotide sequence and scheme of RNA nucleotide pairing in dot-bracket notation of small T-antigen of SV40 virus. Scheme of pairing was obtained with the use of RNA folding program on the site RNAfold WebServer (Institute for Theoretical Chemistry/University of Vienna).

Parameters of pairing scheme: number of nucleotides—522; number of single nucleotides—222; number of nucleotide pairs—150.

Figure 5.10 gives RNA structure of small T-antigen of SV40 virus.

Parameters of RNA structure of small T-antigen of SV40 virus: number of single nucleotides **A**(104), **C**(17), **U**(59), **G**(42) and number of nucleotide pairs **UA**(34), **CG**(29), **AU**(31), **GC**(30), **GU**(13), **UG**(13) are given in parentheses.

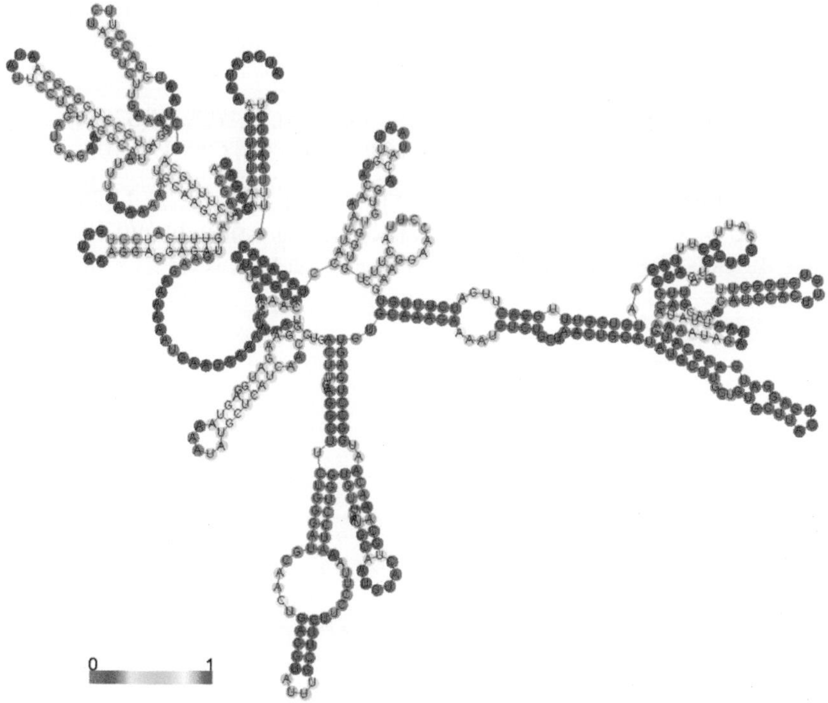

Figure 5.10 RNA structure of small T-antigen of SV40 virus, which is a realization of Clifford algebra $Cl(10, 0)$. The structure was obtained with the use of RNA folding program on the site RNAfold WebServer (Institute for Theoretical Chemistry/University of Vienna). MFE structure drawing encoding base-pair probabilities.

Estimated dimension of Clifford algebra is equal to $4\times(222 + 150) = 1488$. Nearest Clifford algebra $Cl(10, 0) = [528+\ 496-]$ has dimension 1024.

Suppose that RNA nucleotides and nucleotide pairs, which are elements of Clifford algebra, associate with four γ-matrices of quasi-particle (fermion), but which in gauge are defined by the numbers $n_1 \ldots n_{10}$. Excessive RNA nucleotides and nucleotide pairs, which are not elements of Clifford algebra, are identified as neutral.

Clifford algebra elements—RNA nucleotide or nucleotide pair—are quasi-particles (*CL* particles) homomorphic to electron or positron. Square of *CL* particle charge may be positive or negative but it is multiple of electron charge square.

tRNA Clifford algebra, equal to $Cl(6) = Cl(3) \times Cl(3)$, consists of nucleotides and their pairs that may be defined as pairs of *CL* particles. In tRNA there are also neutral nucleotides.

Precise arrangement of Clifford algebra elements in RNA is not known. Among multiple versions we chose following realization of Clifford algebra $Cl(10, 0)$:

$$n_1(80) = \mathbf{A};\ n_8(23) = \mathbf{CG};\ n_{432}(83) = \{\mathbf{AU}(25);\ \mathbf{UA}(25);\ \mathbf{U}(33)\};$$
$$n_{9,10}(20) = \{\mathbf{GU}(10);\ \mathbf{UG}(10)\};\ n_{765}(50) = \{\mathbf{GC}(15);\ \mathbf{G}(25);\ \mathbf{C}(10)\}.$$

The number of neutral nucleotides is equal to 158: **A**(24), **U**(26), **C**(7), **G**(17), **AU**(6), **UA**(9), **GC**(15), **CG**(6), **GU**(3), **UG**(3). Arrangement of neutral nucleotides is also unknown.

APPENDIX 5.2

CLIFFORD ALGEBRAS OF GENETIC CODE AMINO ACIDS

According to different criteria, sets of free amino acids and sets of genetic code amino acid radicals generate opposite Clifford algebras with accuracy within isomorphism (Table 5.3). Aminoacyl-tRNA synthetases (Chapter 2) divide free amino acids into two classes Ψ_I and Ψ_{II} each of 10 amino acids initiating Clifford algebra of Ψ_I and Ψ_{II} sets equal to $Cl(10, 10) \cong Cl(11, 9)$. But, by the number of hydrogen bonds of central (codon-anticodon) pair of mRNA-tRNA pairing, according to genetic code, set of free amino acids generates Clifford algebra $Cl(8, 12) \cong Cl(9, 11)$.

Table 5.3 Clifford algebras of sets of genetic code amino acids.

Amino acid Am	Protonic charge of radical Q_p(Amr)	Q_p(Amr) *mod* 4 $Cl(7, 13)$	#Q_p(Amr) $Cl(11, 9)$	Degeneracy number χ (Am)	#χ (Am) $Cl(12, 8)$	Sets Ψ_I, Ψ_{II} $Cl(10, 10)$	Number of hydrogen bonds $Cl(8, 12)$
A	9	+1	2	4	2	Ψ_{II}	**3**
C	25	+1	2	2	**1**	$\boldsymbol{\Psi_I}$	**3**
D	31	**–1**	**1**	2	**1**	Ψ_{II}	2
E	39	**–1**	2	2	**1**	$\boldsymbol{\Psi_I}$	2
F	49	+1	2	2	**1**	Ψ_{II}	2
G	1	+1	**1**	4	2	Ψ_{II}	**3**
H	43	**–1**	**1**	2	**1**	Ψ_{II}	2
I	33	+1	**1**	3	**1**	$\boldsymbol{\Psi_I}$	2
K	41	+1	**1**	2	**1**	Ψ_{II}	2
L	33	+1	**1**	6	2	$\boldsymbol{\Psi_I}$	2
M	41	+1	**1**	1	**1**	$\boldsymbol{\Psi_I}$	2
N	31	**–1**	**1**	2	**1**	Ψ_{II}	2
P	23	**–1**	**1**	4	2	Ψ_{II}	**3**
Q	39	**–1**	2	2	**1**	$\boldsymbol{\Psi_I}$	2
R	55	**–1**	2	6	2	$\boldsymbol{\Psi_I}$	**3**
S	17	+1	**1**	6	2	Ψ_{II}	**3**
T	25	+1	2	4	2	Ψ_{II}	**3**
V	25	+1	2	4	2	$\boldsymbol{\Psi_I}$	2
W	73	+1	**1**	1	**1**	$\boldsymbol{\Psi_I}$	**3**
Y	57	+1	2	2	**1**	$\boldsymbol{\Psi_I}$	2

Note: Parameters of amino acids, determining positive part of Clifford algebra signature, are enhanced.

Amino acid radicals and their pairs generate, accordingly, Clifford algebras *Cl*(7, 13) and *Cl*(14, 6) isomorphic to algebra *Cl*(11, 9). At quantity #N of different finite groups of N order, amino acid radicals Amr and degeneration numbers χ(Am) of genetic code amino acids divide amino acids into sets with opposite Clifford algebras: *Cl*(11, 9) and *Cl*(12, 8) ≅ *Cl*(9, 11), accordingly.

All protonic charges of amino acid radicals satisfy the condition $\#Q_p(\text{Amr}) = 1, 2$. Among amino acid radicals there are no amino acids with $Q_p(\text{Amr}) = 27, 63$, for which this condition is violated. However, protonic charge $Q_p(\text{Amr}) = 27$ requires the presence of double of triple carbonic bonds in linear radicals of amino acids. Protonic charge $Q_p(\text{Amr}) = 63$ may be realized by elongation of arginine radical chain with $\mathbf{CH_2}$ group, which will result in emergence of additional hydrogen bonds in protein molecule. Radicals of genetic code amino acids seldom form hydrogen bonds with backbone of polypeptide chain. Other odd proton charges, up to 73, satisfy the condition $\#Q_p(\text{Amr}) = 1, 2$.

The most accurate parameter in Table 5.3 is the number of hydrogen bonds of central pair of mRNA-tRNA codon-anticodon pairing. Exactly, mRNA-tRNA pairing results in transformation of genetic information into Clifford algebra *Cl*(8, 12) ≅ *Cl*(12, 8) of free genetic code amino acids: each amino acid has the same number of hydrogen bonds of central pair of mRNA-tRNA codon-anticodon pairing. Besides, in spite of large freedom of choice, the numbers of degeneration of genetic code amino acids are also consistent with Clifford algebra *Cl*(12, 8) of free amino acids.

Interaction of tRNA, aminoacyl-tRNA synthetases and free amino acids conforms to Clifford algebras *Cl*(10, 10), *Cl*(10, 10) and *Cl*(12, 8), accordingly, which contain reverse number of Lorenz intervals [3+ 1–] и [1+ 3–].

Further, we will consider interconnection between Clifford and Lie algebras that are generated by CL particles. It is clear that CL particles are concentrated in atom nuclei and are associated with protons.

Product of CL particle charges with self-action induces chains of algebraic transformation:

$$Z_i \cdot Z_k \begin{cases} \nearrow g_{ik} \longrightarrow \gamma_i\gamma_k + \gamma_k\gamma_i \longrightarrow \gamma_{ik} \\ \searrow C_{ik}^{\ j}\, u_j \longrightarrow u_i u_k - u_k u_i \longrightarrow u_i \wedge u_k \end{cases}$$

$$Z_i \bullet Z_k \bullet Z_j \begin{cases} \nearrow \ \gamma_i\gamma_k\gamma_j + \gamma_j\gamma_i\gamma_k \longrightarrow \gamma_{ikj} \longrightarrow \lambda_{[ikjs]}x_s \\ \searrow \ u_i u_k u_j - u_j u_i u_k \longrightarrow C_{ik}{}^{t} C_{tj}{}^{s} u_s \longrightarrow R_{ikj}{}^{s} u_s \end{cases}$$

where g_{ik} is metric tensor of Clifford algebra;

γ_i—gamma-matrices of Clifford algebra;
u_i—vector field of Lie algebra;
$u_i \wedge u_k$—exterior product in Grassmann algebra;
$C_{ik}{}^{j}$—local structural constants of Lie algebra;
$\lambda_{[ikjs]}$—collecting function of four-dimensional coordinates x_s;
$R_{ikj}{}^{s}$—curvature tensor of four-dimensional Riemann space.

Therefore, product of CL particle charges simultaneously generates gamma-matrices of Clifford algebra and vector field of Lie algebra.

Clifford algebra $Cl(p, q)$ is constructed on the atoms of the same molecule or on the atoms of different but interacting molecules. Positive and negative parts of Clifford algebra signature are defined by the number of protons in atom nuclei. Square of proton charge is a real number equal to the square of electron charge. But proton is a complex particle composed of quarks, and, as every complex particle, may be in an excited state. Proton in excited state possesses imaginary charge. CL particle has two states: common proton with positive charge square and excited proton with negative charge square. In special cases, neutron can also be in the state of CL particle with imaginary charge equal to proton charge, when metrics of mRNA translation interval are generated (Chapter 3). Neutrons are, possibly, somehow connected with Grassmann algebra.

CL particles are factor-particles of quarks of $p - n$ pair. At assumption that d-quark may be divided into two partons and u-quark may not, formation of CL particles may be presented with the following diagram:

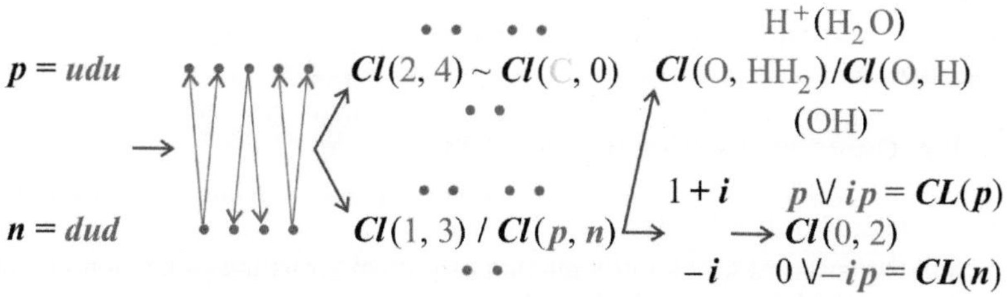

where $Cl(2, 4)$ is Clifford algebra of electron shell of carbon atom.

Selected Bibliography

Stcherbic, V.V., Buchatsky, L.P. Group of alternatives and splicing phenomenon. Problems of ecological and medical genetics and clinical Immunology 115: 30–38, 2013 (in Russian).

Stcherbic, V.V., Buchatsky, L.P. Discrete symmetry of globular proteins. Scientific Transactions of Ternopil National Pedagogical University. Ser. Biol. 46: 106–109, 2011 (in Ukrainian).

RNA and protein splicing

Singer, M., Berg, P. Genes and genomes. California: University Science Books, 1991.

Weaver, R.F. Molecular biology. New York: McGraw-Hill, 2012.

Watson, J.D., Baker, T.A., Bell, S.P., Gann, A., Levine, M., Losick, R. Molecular biology of the gene. New York: Cold Spring Harbor Laboratory, 2004.

Cech, T.R., Zaug, A.G., Grabowski, P.J. *In vitro* splicing of the ribosomal RNA precursor of Tetrahymena: involment of a guanosine nucleotide in the excision of the intervening sequence. Cell 27: 487–496, 1981.

McManus, C.J., Gravely, B.R. RNA structure and the mechanisms of alternative splicing. Curr. Opin. Genet. Dev. 21: 373–379, 2011.

Liang, X., Haritan, A., Uliel, S., Michaeli, S. Trans and cis splicing in trypanosamids: mechanism, factors and regulation. Eukar. Cell 2: 830–840, 2003.

Nielsen, H., Johansen, S.D. Group I introns. Moving in new directions. RNA Biology 6: 375–383, 2009.

Nielsen, H., Fiskaa, T., Birgisdottir, A.B., Haugen, H., Einvik, C., Johansen, S. The ability to form full-length intron RNA circles is a general property of nuclear group I introns. RNA 9: 1464–1475, 2003.

Ralph, D., Huang, J., Van der Ploeg, L.H.T. Physical identification of branched intron side-products of splicing in *Trypanosoma brucei*. EMBO J. 7: 2539–2545, 1988.

Clarke, N.D. A proposed mechanism for self-splicing of proteins. Proc. Natl. Acad. Sci. USA 91: 11084–11088, 1994.

Noble, J.C.S., Prives, C., Manley, J.L. Alternative splicing of SV40 early pre-mRNA is determined by branch site selection. Genes & Dev. 2: 1460–1475, 1988.

Kane, P.M., Yamashiro, C.T., Wolczyk, D.F., Neff, N., Goebl, M., Stevens, T.H. Protein splicing converts the yeast TFP1 gene product to the 69-kD subunit of the vacuolar H(+)-adenosine triphosphatase. Science 250: 651–657, 1990.

Protein structure

Banavar, J.R. and Maritan, A. Physics of Proteins. Annu. Rev. Biophys. Biomol. Struct. 36: 261–280, 2007.

Branden, C., Tooze, J. Introduction to protein structure. 2nd ed. New York: Garland Publishing, Inc., 1999.

Rees, A.R., Sternberg, M.J.E. From cells to atoms. An illustrated introduction to molecular biology. Oxford: Blackwell Scientific Pub., 1984.

Berezovsky, I.N. Discrete structure of van der Waals domain in dlobular proteins. Protein Engineering 16: 161–167, 2003.

Lee, B. and Richards, F.M. The interpretation of protein structure: estimation of static accessibility. J. Mol. Biol. 55: 379–400, 1971.

Ramachandran, G.N., Sasisekharan, V. Conformation of polypeptides and proteins. Adv. Prot. Chem. 28: 283–437, 1968.

DSSP algorithm

Martin, J., Letellier, G., Marin, A., Taly, J.-F., de Brevern, A., Gibrat, J.-F. Protein secondary structure assignment revisited: a detailed analysis of different assignment methods. BMC Structure Biology 5: 17, 2005.

Kabsch, W., Sander, C. Dictionary of protein secondary structure: pattern recognition of hydrogen-bonded and geometrical features. Biopolymers 22: 2577–2637, 1983.

Geometry and theory of groups

Dubrovin, B.A., Fomenko, A.T., Novikov, S.P. Modern geometry—methods and applications. Part II. The geometry and topology of manifolds. New York: Springer-Verlag, 1985.

Trofimov, V.V. Introduction to geometry of manifolds with symmetry. Springer Netherlands, 1994.

Coxeter, H.S.M., Moser, W.O.J. Generators and relations for discrete groups. Berlin: Springer-Verlag, 1972.

Petrashen, M.I. and Trifonov, E.D. Applications of group theory in quantum mechanics. Cambridge: M.I.T. Press, 1969.

Vilenkin, N.J. Special functions and the theory of group representations. American Mathematical Society, 1968.

Supplementary Materials

DNA coding of signature of Clifford algebra

Coding DNA chain consists of intervals [3+ 1–]; matrix DNA chain, along which RNA polymerase moves, consists of intervals [1+ 3–]: **T**[– – – +], **A**[– – + –], **G**[– + – –], **C**[+ – – –]. Here a plus sign marks the moment of recognition of adequate nucleotide in four-interval box. Coding and matrix DNA chains are Lorentz intervals of certain Clifford algebra.

The fragment of coding sense DNA, which is the result of transcription, is referred to as a gene. Coding DNA chain does not participate directly in transcription; therefore, coding content of nucleotides in a sense chain may be formulated as serial numbers of some sequence of codes. To be more specific, we will code DNA nucleotides with binary code (DNA_{bin}), which allows 24 code combinations. Such an approach describes coding and matrix DNA chains as sign signatures (charges) of some Clifford algebra $Cl(Q_g)$, where Q_g is the length of gene in DNA nucleotides.

Both coding and matrix DNA chains define six Clifford algebras within the accuracy to sign signature arrangement along DNA chain.

Gene Clifford algebras are defined by pairs of nucleotides along DNA chain: *Cl*(**C**+**A**, **G**+**T**), *Cl*(**T**+**A**, **G**+**C**), *Cl*(**G**+**A**, **C**+**T**), *Cl*(**G**+**C**, **A**+**T**), *Cl*(**G**+**T**, **A**+**C**), *Cl*(**C**+**T**, **A**+**G**), where **C**+**A** indicates the sum of nucleotide numbers in coding chain q(**C**) + q(**A**), etc.

Using transformation rules of Clifford algebras, it is not difficult to demonstrate that there may always be found such code combination for DNA nucleotides wherein Clifford algebras of coding and matrix chains are isomorphic.

As an example, we will consider Clifford algebras that are coded by gene H3F3A of histone H3 (Homo sapiens) (Figure sm1). Numerical parameters of the gene: q(**A**) = 207, q(**G**) = 186, q(**C**) = 179, q(**T**) = 185; Q_g = 757. The gene determines six Clifford algebras:

Cl(386, 371), *Cl*(392, 365), *Cl*(393, 364), *Cl*(365, 392), *Cl*(371, 386), *Cl*(364, 393).

```
GTGTTCGCAGCCGCCGCCGCGCCGCCGTCGCTCTCCAACGCCAGCGCC
GCCTCTCGCTCGCCGAGCTCCAGCCGAAGGAGAAGGGGGGTAAGTAAG
GAGGTCTCTGTACCATGGCTCGTACAAAGCAGACTGCCCGCAAATCGA
CCGGTGGTAAAGCACCCAGGAAGCAACTGGCTACAAAAGCCGCTCGCA
AGAGTGCGCCCTCTACTGGAGGGGTGAAGAAACCTCATCGTTACAGGC
CTGGTACTGTGGCGCTCCGTGAAATTAGACGTTATCAGAAGTCCACTG
AACTTCTGATTCGCAAACTTCCCTTCCAGCGTCTGGTGCGAGAAATTG
CTCAGGACTTTAAAACAGATCTGCGCTTCCAGAGCGCAGCTATCGGTG
CTTTGCAGGAGGCAAGTGAGGCCTATCTGGTTGGCCTTTTTGAAGACA
CCAACCTGTGTGCTATCCATGCCAAACGTGTAACAATTATGCCAAAAG
ACATCCAGCTAGCACGCCGCATACGTGGAGAACGTGCTTAAGAATCCA
CTATGATGGGAAACATTTCATTCTCAAAAAAAAAAAAAAAAATTTCTC
TTCTTCCTGTTATTGGTAGTTCTGAACGTTAGATATTTTTTTTCCATG
GGGTCAAAAGGTACCTAAGTATATGATTGCGAGTGGAAAAATAGGGGA
CAGAAATCAGGTATTGGCAGTTTTTCCATTTTCATTTGTGTGTGAATT
TTTAATATAAATGCGGAGACGTAAAGCATTAATGCAA
```

Figure sm1 DNA nucleotide sequence of the gene H3F3A.

Isomorphism of Clifford algebras for coding and matrix DNA chains is defined by comparing corresponding algebras, e.g., *Cl*(**C**+**A**, **G**+**T**) and *Cl*(**C**+**T**, **G**+**A**), i.e., *Cl*(386, 371) ~ *Cl*(364, 393) for one of 24 coding combinations.

mRNA, coding histone protein H3, contains Clifford algebra

Cl(193, 218), which is subalgebra of algebra *Cl*(386, 371), but in coding combination *Cl*(**G**+**U**, **A**+**C**).

Further, we will consider presentations of Clifford algebras of DNA chains that are determined on the basis of numerical content of individual nucleotides. Each nucleotide N is considered as a quantized surface with distinguished points, which define square matrix $\boldsymbol{M}_N$ of k_N order. Numerical values of $\boldsymbol{M}_N$ matrix elements may be determined, e.g., by Gaussian curvature of the surface. Symmetrical part of $\boldsymbol{M}_N$ matrices has signature (p_N, q_N). Matrix $\boldsymbol{M}_N$ of nucleotide N is supposed to not change along DNA chain.

At present, matrices $\boldsymbol{M}_N$ for DNA nucleotides are unknown. Therefore, instead of matrices $\boldsymbol{M}_N$ we will consider limiting diagonal matrices of nucleotides:

maximal ones, on the basis of structural-vectorial linear field (Chapter 3); minimal ones, on the basis of donor-acceptor hydrogen bonds.

Assume that real M_N matrices have dimension in the interval of dimensions of limiting diagonal matrices.

Proton charges of DNA nucleotides are determined on the basis of their structural-vectorial fields:

$$Q_p(\mathbf{A}) = Q_p\{\mathbf{5C\ 4H\ 5N}\} = 69,\ Q_p(\mathbf{T}) = Q_p\{\mathbf{5C\ 5H\ 2O\ 2N}\} = 65,$$
$$Q_p(\mathbf{G}) = Q_p\{\mathbf{5C\ 4H\ 1O\ 5N}\} = 77,\ Q_p(\mathbf{C}) = Q_p\{\mathbf{4C\ 4H\ 1O\ 3N}\} = 57.$$

Since Clifford algebra $Cl(\mathbf{A}, \mathbf{T}) = Cl(69, 65)$ is isomorphic to Clifford algebra $Cl(\mathbf{G}, \mathbf{C}) = Cl(77, 57)$, but not isomorphic to Clifford algebra $Cl(\mathbf{C}, \mathbf{G}) = Cl(57, 77)$, we will select absolute coding of DNA nucleotides as follows:

in coding chain

$\mathbf{A} = Cl(69, 0)$, $\mathbf{T} = Cl(0, 65)$, $\mathbf{C} = Cl(57, 0)$, $\mathbf{G} = Cl(0, 77)$;

in matrix chain

$\mathbf{T} = Cl(65, 0)$, $\mathbf{A} = Cl(0, 69)$, $\mathbf{G} = Cl(77, 0)$, $\mathbf{C} = Cl(0, 57)$.

In such coding, Clifford algebra of complementary pair **AT** (**CG**) coincides with Clifford algebra of **AT** (**CG**) pair along coding and matrix chains.

Absolute coding of DNA nucleotides determines the signature of maximal Clifford algebra of coding chain.

Gene H3F3A determines the signature of maximal Clifford algebra $Cl(24486, 26347)$.

Clifford algebras of DNA nucleotides on the basis of donor-acceptor hydrogen bonds $\mathbf{A}Cl(2, 3)$, $\mathbf{T}Cl(2, 1)$, $\mathbf{C}Cl(2, 2)$, $\mathbf{G}Cl(3, 3)$ determine the signature of minimal Clifford algebra of coding chain.

Gene H3F3A determines the signature of minimal Clifford algebra $Cl(1700, 1722)$.

Thus, coding DNA chain is really a code of character signature of certain Clifford algebra.

DNA transcription and bundle of Clifford algebra

RNA polymerase, moving along DNA as a particle, forms Lorentz intervals [3+ 1–] independently of transcribed DNA nucleotides. Factor algebras of DNA reading differ for **GC** and **AT** pairs and are equal $Cl(1, 1)$ and $Cl(0, 2)$, accordingly.

Transcription of **GC** pairs, considered as automorphism of Clifford algebra $Cl(1, 1)$, is its trivial bundle, so long as $Cl(1, 1) = [3+\ 1-]$.

For **AT** pair, trivial bundle of Clifford algebra $Cl(0, 2) = [1+\ 3-]$ does not exist. But there is nontrivial bundle $Cl(0, 2)$ with nucleotide substitution $\mathbf{T} \rightarrow \mathbf{U}$.

Reading of nucleotide $\mathbf{G}_{\mathrm{DNA}}$ and $\mathbf{C}_{\mathrm{DNA}}$ in matrix chain is performed by including $\mathbf{C}_{\mathrm{RNA}}$ and $\mathbf{G}_{\mathrm{RNA}}$ nucleotides in RNA chain:

$$\left.\begin{array}{l} \mathrm{G}_{\mathrm{DNA}} \rightarrow \mathrm{C}_{\mathrm{RNA}} \times \boldsymbol{Cl(1, 1)^{+1}} \\ \mathrm{C}_{\mathrm{DNA}} \rightarrow \mathrm{G}_{\mathrm{RNA}} \times \boldsymbol{Cl(1, 1)^{-1}} \end{array}\right\rangle \boldsymbol{Cl(1, 1)} \xrightarrow{+1,\,+1} \boldsymbol{Cl(2, 0)_{\mathrm{POL}}}$$

In this formula, charges ± 1 determine trivial bundle of algebra $Cl(1, 1)$, while charges $(+1, +1)$ determine algebra of RNA polymerase movement $Cl(2, 0) = [3+\ 1-]$.

Reading of nucleotide $\mathbf{A}_{\mathrm{DNA}}$ and $\mathbf{T}_{\mathrm{DNA}}$ in matrix chain is performed by including $\mathbf{U}_{\mathrm{RNA}}$ and $\mathbf{A}_{\mathrm{RNA}}$ nucleotides in RNA chain:

$$\left.\begin{array}{l} \mathbf{T}_{\mathrm{DNA}} \rightarrow \mathbf{A}_{\mathrm{RNA}} \times \boldsymbol{Cl(0, 2)^{-1}} \\ \mathbf{A}_{\mathrm{DNA}} \rightarrow \mathbf{U}_{\mathrm{RNA}} \times \boldsymbol{Cl(0, 2)^{-1}} \end{array}\right\rangle \boldsymbol{Cl(0, 2)} \xrightarrow{+1,\,-1} Cl(1, 1)_{\mathrm{POL}}$$

In this formula, charges $(-1, -1)$ determine nontrivial bundle of algebra $Cl(0, 2)$, while charges $(+1, -1)$ determine algebra of RNA polymerase movement $Cl(1, 1) = [3+\ 1-]$.

Butterfly Spin(*p*, *q*)-structure

The group Spin(p, q) is known to cover twice the group $SO(p)\times SO(q)$. As opposed to simple Lie groups, spinor groups act on geometric structures simultaneously both on the left and on the right. So called adjoint representation of the spinor group $T(\boldsymbol{r})$ is built as reflection of the vector $\boldsymbol{r}$ relative to a hyperplane perpendicular to the vector $\boldsymbol{a}$:

$T(\boldsymbol{r}) = \boldsymbol{ara}^{-1}$. Vector $\boldsymbol{r}$ consists of linear combination of basis vectors of Clifford algebra; vector $\boldsymbol{a}$ represents even product components of basis vectors of Clifford algebra. The group Spin(p, q) covers twice *specularly nontrivially* the group $SO(p)\times SO(q)$. Meaning of nontrivial covering is that the intersection $SO(p)\cap SO(q)$ has antisymmetric structure and binds between themselves the regions coded by the signature (p, q); accordingly, the group Spin(p, q) becomes the simply connected group. At action of the group Spin(p, q) on geometrical structures, the region of intersection

SO(p)∩SO(q) becomes neutralized and the group SO(p, q) presents itself as SO(p)×SO(q). Intersection SO(p)∩SO(q) induces Clifford algebra $Cl(p-q, p-q)$, $p \geq q \geq 1$, so long as the region $(p-q)$ is paired. If $p = q$, antisymmetrical structure is absent.

Generally, every geometrical structure may be mapped onto some matrix with arbitrary structure, but symmetric part of the matrix determines bundle of the bases, i.e., generating vectors of Clifford algebra; nonsymmetrical part of the matrix determines Clifford algebra of nontrivial design of intersection region in the Spin structure.

Naturally, dual presentation of the group Spin(p, q) is realized (Figure sm2):

$$\mathrm{T}(\boldsymbol{r}) = \boldsymbol{ara} = \boldsymbol{r}^{\dagger},$$

where $\boldsymbol{r}$ are even components of Clifford algebra;

$\boldsymbol{r}^{\dagger}$ are reversing even components of Clifford algebra;

$\boldsymbol{a}$ are basic vectors of Clifford algebra.

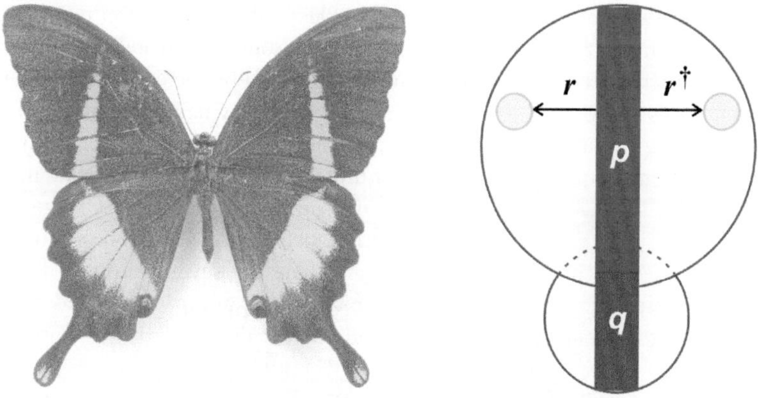

Figure sm2 Dual presentation of the group Spin(p, q).
(**Left**) Butterfly. (**Right**) Mathematical model of the butterfly wings.

Figure sm2 shows mathematical model of the butterfly wings. Regions of basic vectors are simplified; actually, they are strongly mixed up.

If even components of Clifford algebra do not intersect with basic vectors, p-region parametrizes components of 0, 4, 8... power and q-region parametrizes components of 2, 6, 10 ... power.

If even components of Clifford algebra intersect with basic vectors, antisymmetrical structure is formed.

For instance, $e_1^2 = +1, e_2^2 = -1$. Then

$$e_1(e_{1234})e_1 = e_{2341} = -e_{4321};\ e_2(e_{1234})e_2 = -e_{2134} = e_{4321}.$$

Geometrical structure of regions of the group Spin(p, q) is rather composite, which can be seen in Figure sm2. One can suppose that Clifford algebra components of the same power form domains.

Lie algebra SO(q) $\subset$ SO(p), which is realized on second power components of Clifford algebra $Cl(p, q)$, defines synchronous motion of fore and hind winds of the butterfly.

Centrosome Spin$^{\Psi}$(p, p)-structure

Centrosomes are organelles present in most cells of vertebrates and lower plants. They consist of two centrioles and pericentriolar material. Centrioles are symmetric cylindrical structures with small volume (1 μ^3) that are located almost in the centre of the cell perpendicularly to one another (Figure sm3).

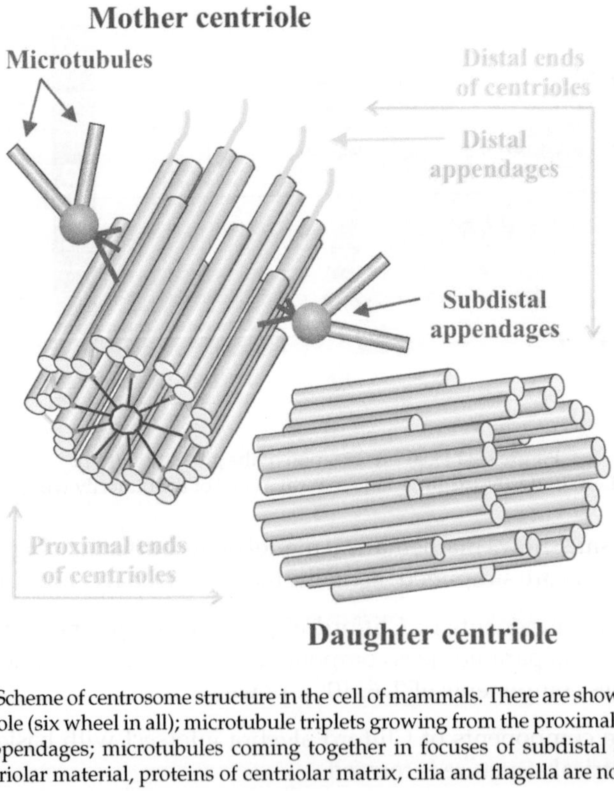

Figure sm3 Scheme of centrosome structure in the cell of mammals. There are shown: cartwheel of procentriole (six wheel in all); microtubule triplets growing from the proximal end; mother centriole appendages; microtubules coming together in focuses of subdistal appendages. Pericentriolar material, proteins of centriolar matrix, cilia and flagella are not shown.

Centrioles are surrounded by amorphic fibrillar mass (matrix), from which microtubules are stepping aside. Centrioles have point central symmetry of the 9th order, which is conditioned by nine sets of microtubule triplets (designated as ABC) that are just forming a cylinder. All microtubule triplets are oriented in parallel and inclined equally towards the cylinder circle at an angle of 45°. The cylinder is 0.2 μm in diameter and its length is 0.3–0.5 μm. All microtubules have the same structure and are formed from heterodimer of α-, β-tubulins. Each microtubule consists of fibers (protofilaments), whose numbers are usually equal to 13, 10, 10.

Centrosome centrioles are in close contact through proximal ends. One centriole (mother) has additional appendages and pericentric satellites on its distal end. The other one (daughter) does not have additional structures.

Pericentriolar material, surrounding centrioles, is the area of microtubule growth. Each microtubule grows on the basis of circular multiprotein complex containing molecules of γ-tubulin, which is similar by its structure with α- and β-tubulin. Matrix elements are a basis for fixation of proteins that take part in the growth of microtubules, appendages and microfilaments that connect centrioles.

In certain cases, microtubule triplet is substituted for one or two microtubules, when the centriole (basal body) is the growth centre for cilia or flagella.

Centrosome is replicated in each cell cycle. During duplication process in G1/S phase of cell cycle, many centriolar and pericentriolar components accumulate and centrosome is replicated via separation into two centrioles. Each daughter centriole grows perpendicularly to the mother centriole. After replication in prophase of cell cycle, centrosomes became situated at opposite poles of the cell mitotic spindle. Centrioles can also arise from amorphous cytoplasmic material, in which initial structure-forming matrices are not observed.

Centrosome performs different functions in the cell. Two centrosomes form bipolar mitotic spindle, necessary for cleavage of two daughter chromosomes and their separation on the opposite poles of the cell. Centrosome is an organizing centre of pericentriolar material. Centrosome proteins ensure the growth of microtubules.

Suppose that centrosome is the presentation of some Clifford algebra $Cl(p, p)$. Let us build symmetric matrix with signature (p, p).

We will present a section of the space structure as a quantized surface with distinguished points that determine numerical values of a square matrix M of k order.

It is supposed that the matrix M, like the quantized surface, is asymmetric. Let us decompose the matrix M on the sum of symmetric A$\{a, a\}$ and skewsymmetric B$\{-b, b\}$ matrices:

$$M = \begin{vmatrix} \ddots & a \\ a & \ddots \end{vmatrix} + \begin{vmatrix} \diagdown & b \\ -b & \diagdown \end{vmatrix}$$

Complex presentation of the matrix M determines symmetric matrix with signature (p, p):

$$M_c = \begin{vmatrix} \ddots & a \\ a & \ddots \end{vmatrix} \otimes \begin{vmatrix} -1 & \diagup \\ \diagup & 1 \end{vmatrix} + \begin{vmatrix} \diagdown & b \\ -b & \diagdown \end{vmatrix} \otimes \begin{vmatrix} \diagdown & 1 \\ -1 & \diagdown \end{vmatrix} = \begin{array}{cc} \begin{vmatrix} \ddots & -a \\ -a & \ddots \end{vmatrix} & \begin{vmatrix} \diagdown & b \\ -b & \diagdown \end{vmatrix} \\ \begin{vmatrix} \diagdown & -b \\ b & \diagdown \end{vmatrix} & \begin{vmatrix} \ddots & a \\ a & \ddots \end{vmatrix} \end{array}$$

Let us present centrosome as the sum of two matrices—mother M_m and daughter M_d ones:

$$M_m = \begin{array}{c} \begin{vmatrix} \diagdown & b \\ -b & \diagdown \end{vmatrix} \\ \begin{vmatrix} \ddots & a \\ a & \ddots \end{vmatrix} \end{array}; \quad M_d = \begin{vmatrix} \ddots & -a \\ -a & \ddots \end{vmatrix} \begin{vmatrix} \diagdown & -b \\ b & \diagdown \end{vmatrix}$$

It is known from experiments that the growth of new centrioles occurs from proximal ends of old centrioles. Mother and daughter centrioles differ from the side of distal ends; therefore, symmetric and skewsymmetric parts of matrices M_m and M_d will correspond to proximal and distal ends of centrioles, accordingly.

Centrosome replication results in formation of two new pairs of centrioles:

$$M_{md} = \begin{array}{ccc} \begin{vmatrix} \diagdown & b \\ -b & \diagdown \end{vmatrix} & & \\ \begin{vmatrix} \ddots & a \\ a & \ddots \end{vmatrix} & \begin{vmatrix} \ddots & -a \\ -a & \ddots \end{vmatrix} & \begin{vmatrix} \diagdown & -b \\ b & \diagdown \end{vmatrix} \end{array}; \quad M_{dd} = \begin{array}{cc} \begin{vmatrix} \diagdown & b \\ -b & \diagdown \end{vmatrix} & \\ \begin{vmatrix} \ddots & a \\ a & \ddots \end{vmatrix} & \\ \begin{vmatrix} \ddots & -a \\ -a & \ddots \end{vmatrix} & \begin{vmatrix} \diagdown & -b \\ b & \diagdown \end{vmatrix} \end{array}$$

New pair of centrioles M_{dd} rotates by the angle π (with transformation $a \leftrightarrow -a$) and they become located on the opposite pole of the cell mitotic spindle:

$M_{dd}^{\pi} =$ 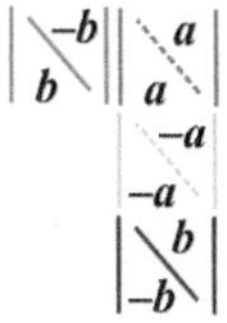

Covering matrix of the spinor group is:

$M_{md, dd}^{\pi} =$ 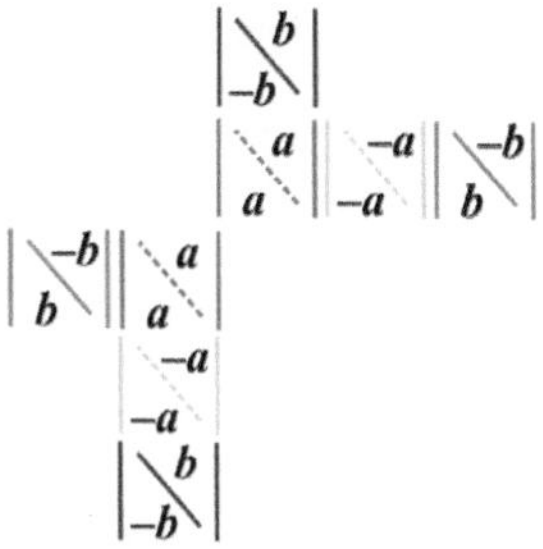

Matrix $M_{md', dd}^{\pi}$ nontrivially twice cover the matrix M_c.

Thus, Clifford algebra $Cl(p, p)$ defines not only the structure of centrosome but also contains the spinor group Spin(p, p) that determines the centrosome replication.

Selected readings

Azimzadeh, J., Marshall, W.F. Building the centriole. Curr. Biol. 20: R816–R825, 2010.

Bornens, M. The centrosome in cells and organisms. Science 335: 422–426, 2012.

Index

For Product Safety Concerns and Information please contact our EU representative GPSR@taylorandfrancis.com Taylor & Francis Verlag GmbH, Kaufingerstraße 24, 80331 München, Germany

Printed and bound by CPI Group (UK) Ltd, Croydon, CR0 4YY
01/05/2025
01858583-0001